Praise for
The Year-round Solar Greenhouse

Particularly for aquaponics growers, solar greenhouse design is a no-brainer. Schiller and Plinke have created a practical, easy to read guide that enables anyone to design and build their own sustainable, year-round greenhouse. I highly recommend it for aquaponic growers, and any gardener looking to extend their season.

—Sylvia Bernstein, author, *Aquaponic Gardening*

Schiller's book is an important resource that will help farmers and greenhouse operators leverage innovation for sustainable and profitable food production. I believe that agricultural innovation for economic and ecological sustainability is the most important opportunity facing humanity, and this book should be part of your tool kit.

—Gaelan Brown, author, *The Compost-Powered Water Heater*

Year-round food production in the emerging post carbon society will require solar greenhouses at many scales. Whether an attached home greenhouse or large commercial bioshelter, successful long term food production in these bio-structures requires artful design, careful planning, quality construction and carefully integrated systems of light, heat, ventilation and well managed growing spaces. Lindsey Schiller and Mark Plinke have provided an essential tool to ensure success in all these areas. This book is extensively researched, written with personal experience and full of essential facts and figures rendered simple and accessible.

—Darrell E. Frey, author, *The Bioshelter Market Garden*

The Year-Round Solar Greenhouse is an invaluable primer for anyone looking to grow their own food year-round without fossil fuels. The science and methods are well explained, meticulously documented, and easy to understand. A great resource!

—Dan Chiras, author, *Power from the Sun, Chinese Greenhouses,* and
The Homeowner's Guide to Renewable Energy

Well researched and thorough, it's a contribution of her effort to convey to us all the information on the subject. The author is educated and writes in a manner easily understood, and to the point. She has done us all a favor with this book. Each chapter ends with a summary "takeaways" that gives extra reading references, books, CD's etc. covering all related topics for whatever your particular need may be. What you need to learn about solar green houses you WILL find by starting with this book!

—Leslie Jackson, co-author, *Rocket Mass Heaters*

THE
YEAR-ROUND
Solar Greenhouse

How to Design and Build a Net-Zero Energy Greenhouse

LINDSEY SCHILLER *with* **MARC PLINKE**

new society
PUBLISHERS

Cover design by Diane McIntosh.
Cover photos: bottom left © Penn and Cord Parmenter, Smart Greenhouses LLC
Top photo of Golden Hoof Farm greenhouse, © Lindsey Schiller/
Ceres Greenhouse Solutions
Sun element © iStock
All others © Lindsey Schiller/Ceres Greenhouse Solutions

Printed in Canada. First printing October 2016.

Inquiries regarding requests to reprint all or part of *The Year-Round Solar
Greenhouse* should be addressed to New Society Publishers at the address below.
To order directly from the publishers, please call toll-free (North America)
1-800-567-6772, or order online at www.newsociety.com
Any other inquiries can be directed by mail to:

New Society Publishers
P.O. Box 189, Gabriola Island, BC V0R 1X0, Canada
(250) 247-9737

LIBRARY AND ARCHIVES CANADA CATALOGUING IN PUBLICATION

Schiller, Lindsey, 1987-, author
The year-round solar greenhouse : how to design and build a net-zero
energy greenhouse / Lindsey Schiller with Marc Plinke.

Includes index.
Issued in print and electronic formats.

ISBN 978-0-86571-824-1 (paperback).—ISBN 978-1-55092-618-7 (ebook)

1. Solar greenhouses. 2. Solar greenhouses—Design and construction.
3. Solar greenhouses—Heating and ventilation—Handbooks, manuals, etc.
4. Greenhouse gardening. 5. Solar energy—Passive systems. I. Plinke, Marc,
author II. Title.

SB415.S35 2016   690'.8924   C2016-905437-3
C2016-905438-1

Funded by the Government of Canada | Financé par le gouvernement du Canada | Canada

New Society Publishers' mission is to publish books that contribute in fundamental
ways to building an ecologically sustainable and just society, and to do so with the least
possible impact upon the environment, in a manner that models that vision.

Contents

Section IV: Growing in the Greenhouse

Introduction

The snow is shin deep, the mercury well below freezing. In the stunning clarity of winter sunshine, a complex triangle of glass rises from among the dazzling white drifts. A layer of condensation obscures the details of the verdant world inside, but as I draw closer, the green takes shape: a forest of kale, hanging baskets of alyssum, beguiling arch of pole beans. Hyacinths float atop vats of greenish water, as catfish swim in lazy circles.

In contrast with the cold, white world I just stepped out of, this winter landscape feels like paradise. As I quickly shed layers, my muscles release their frigid tension and my face relaxes into a smile. The air is humid, teeming with the sweet smell of soil, of respiring plants, of life.

—Elise Hugus,
"The Cape Cod Ark: A Study in Self-Sufficiency,"
Edible Cape Cod
Winter, 2014.

In the winter of 2011, I went out to see a bizarre-looking structure on a farm in East Boulder, Colorado. The building was a prototype net-zero-energy greenhouse funded by the Colorado Department of Agriculture, but except for some glass, it bore little resemblance to a greenhouse. Wood boards, acting as light reflectors, protruded from the front. A sharply peaked sawtooth roof reflected light down to the plants inside. Moveable boards of insulation opened and lowered between panes of windows.

Standing in the humid room teeming with vegetables, I saw a spectacled man shuffling along the wall. He occasionally stopped to plug in his Macbook into various gadgets. "He's just getting data," the tour guide explained with a wave. It was a brief moment but one I remember well—as a marker when life took a different direction.

The data-collecting gentleman was Marc Plinke, who would turn out to be my business partner and co-author. In a follow-up meeting over coffee, we discussed the experimental greenhouse, the potential for the design and technology, and future business aspirations. We were in the same mindset, and a few months later we started a business to test and refine the concepts and make energy-efficient, sustainable greenhouses available to a wider market. We named it *Ceres Greenhouse Solutions* after the Roman goddess of agriculture, inspired by an image of an unruly-haired goddess I saw a few months earlier on European currency. As a business, Ceres has provided an incredible vehicle in which to research and apply new ideas to net-zero-energy greenhouses. It has been an incubator that allowed us to tweak and improve designs with every iteration, exploring and developing new ways to store and transfer the heat of a greenhouse.

Much of our development effort went into a type of ground-to-air heat exchanger we call a *Ground-to-Air Heat Transfer* (GAHT) system, which stores excess daytime heat from the greenhouse in the soil underground. We've worked to simplify the GAHT system, making it more cost-effective and easy to install and recorded huge quantities of data to improve their efficiency. We've designed and installed hundreds of GAHT systems all over the world (from Sweden to Brunei), in growing operations large and small, demonstrating the universal applicability of the system.

We also design and build greenhouses themselves, concentrating on well-insulated structures for growers in harsh climates where the growing season is limited to a few frost-free weeks. Our primary greenhouse design is a shed-style structure with a polycarbonate roof and glass view windows or polycarbonate walls. (The specifics of glazing, angles, dimensions, etc. are customized to meet the grower's needs and location.)

This arrangement works well to maximize year-round light and retain heat for harsh climates.

The design is efficient, attractive and cost-effective for a range of growers. We've had the opportunity to work with backyard gardeners, schools, farms and high-tech commercial greenhouses. The potential for year-round, sustainable growing environments spans all these sectors— plus retirement homes, office buildings, prisons, hospitals...wherever there is space and a need for food.

Our Approach

An internet search for "greenhouse" yields an array of companies that have "the best" greenhouse design. The best materials, light transmission, durability...whatever it is, many claim to have the sole superior greenhouse. Hopefully, you already know to take these statements with a grain of salt. The truth is there is no one "right" greenhouse design; the best greenhouse for you depends on your climate, your goals, and your budget. Texas has a very different climate than Maine. Both can have highly functional, energy-efficient greenhouses, but they require different solutions.

A solar greenhouse is a particular type of greenhouse. It relies on a tailored approach for the creation of a structure that works with the local climate and resources, using the sun as the predominant energy source not only for growth but for the structure's energy needs. The aim of this book is to explain the array of options available for designing and building abundant, year-round greenhouses. Moreover, it serves to provide an explanation of the fundamental concepts that allow solar greenhouses to work, so that you can navigate the choices out there and find an approach that is truly right for your situation.

The current literature on solar greenhouses consists of books that are either very dated (from the 1970s) or that describe a very specific building method that the author has adopted. You can find a book on building a greenhouse out of recycled tires, or with a Chinese design, or underground. We wanted to create a resource that fairly compares and contrasts all these approaches, and explains the fundamental concepts behind them, so that you can go on to create a year-round oasis that is truly right for you.

By choosing to look at the fuller picture, we can't explain the details of each system, such as how to build a build a rocket mass heater or install a solar panel system. These topics deserve in-depth discussion, but we could not write a tome and had to draw lines somewhere. Thus, we provide an overview of systems and building methods, and conclude most chapters with recommendations for further reading. We did not have the space to provide step-by-step building instructions for every construction type, but provide the resources that do, so you can go on to take the next step.

Solar greenhouse design is unique in that it stands at the intersection of simple, time-tested methods and advanced technologies. Heating methods can be a few drums of water or an intricate solar hot water system. The range of technology makes costs for energy-efficient greenhouses extremely wide-ranging, from less than $1 per square foot to over $100. We provide ballpark figures in order to help you make decisions, but leave the final budgeting work to you, recognizing that costs vary greatly by location and are always changing.

How This Book Is Structured

The book begins with the big picture: what is a solar greenhouse and where does it fit in in the range of growing options? It ends with the final step of the design process, how to integrate different growing methods and laying out the greenhouse floor and planting plan.

The heart of this book lies in the middle two sections. Section 2 explains the fundamentals of solar greenhouse design. It concentrates on the greenhouse structure—the glazing, insulation, ventilation, overall design, and construction options. Section 3 describes the range of sustainable heating and cooling options for greenhouses. These methods span active (electric) and passive methods, and range in complexity. Most focus on ways to store the excess heat during the day in the greenhouse for use as heating at night, allowing the greenhouse to become "self-heating," a phrase we often use to differentiate our design philosophy from more petroleum-dependent approaches. The greenhouse, abiding by the greenhouse effect, creates all the heat needed. The "heating and cooling systems" we describe merely transfer that heat in simple and elegant ways.

Every chapter in this book includes examples of successful solar greenhouse projects. Designed by researchers, experts, backyard tinkerers and home growers, these examples are meant to exapnd the possibilities, while highlighting a unique system or design feature. Perhaps it's a rocket mass heater combined with a hot tub, or a solar hot water system integrated with an aquaponics greenhouse. The short profiles of successful greenhouses aim to inspire and guide you in creating a productive, year-round greenhouse unique to you.

Acknowledgments

Writing a book can be an arduous, often stressful task. Fortunately, I had many people who made the road easier, contributing their time, ideas and expertise. Thank you to the numerous growers, experimenters, researchers and backyard innovators who contributed to this book by sharing their experiences and photos. Moreover, it is through their work and those like them, that solar greenhouse design is what it is today. In particular, thank you to these innovators:

- Earle Barnhart and Hilde Maingay, The Green Center
- JD and Tawnya Sawyer, Colorado Aquaponics and Flourish Farms
- Alice and Karel Starek, The Golden Hoof Farm
- Shane Smith, Cheyenne Botanic Gardens
- Gaelan Brown, Agrilab Technologies
- Rob and Michelle Avis, Verge Permaculture
- Penn and Cord Parmenter, Smart Greenhouses LLC
- L. David Roper
- Amory Lovins and the staff at Rocky Mountain Institute
- Susanna Raeven, Raven Crest Botanicals
- Dan Chiras, The Evergreen Institute

- Will Allen and the staff at Growing Power
- David Baylon, Ecotope, Inc.
- The staff at Jasper Hill Farm

I want to sincerely thank my business partner/co-author, Marc Plinke, for his time and support in this endeavor, as well as his continual thoughtfulness and openness to new ideas. Marc has managed to simultaneously run a rapidly growing business and help write a book, all while still making time for his amazing family. For that, I am indebted to them all.

Additionally, many thanks to:

- My impromptu editor, Alex Kalayjian, for making this journey much more enjoyable and being ruthless with a fine-tipped editing pen.
- Caleb Rockenbaugh of Insideo Collective, for enhancing this book with energy models and quantitative recommendations, putting curiosity over compensation, and believing in the giving economy.
- The friends and family who contributed in ways large and small, reading chapters, donating input based on their expertise, and simply asking "how's it going?"
- The staff and editors at New Society, for taking this project on and their commitment to sustainable business practices.
- Josh Holleb and Karin Uhlig, for making Ceres, and thus this book, possible.
- A dog named Charlie, whose incessant desire for walks kept me sane.

Lastly and above all, thank you to George and Joan Schiller, whose incredible love and support has opened every door I walk through.

THE BIG PICTURE

CHAPTER 1

What Is a Solar Greenhouse?

"Don't all greenhouses use the sun?"
"You mean a greenhouse with solar panels?"

Stand in front of a sign that says "Solar Greenhouses" at a green products trade show, and you'll frequently be asked these questions. Few people are familiar with the concept of a solar greenhouse. Simply put, it's a greenhouse that uses the sun's energy not only for growth, but also for passive heating; thus, it is able to maintain suitable growing temperatures without reliance on fossil fuels.

Indeed, all greenhouses use the sun for heat and growth *during the day*. At night, most greenhouses quickly lose all that heat due to the poor insulating quality of their materials. On a winter morning, a standard unheated greenhouse usually is only a few degrees warmer (if at all) than the outdoor temperature. Moreover, unless it is ventilated or artificially cooled, a standard greenhouse traps so much heat during the day that it will drastically overheat.

Energy author Dan Chiras once used an excellent analogy to give a quick picture of traditional greenhouses: Imagine living in a tent.[1] When it's 90°F (32°C) outside, sitting in a closed tent is the last place you want to be. When it's 32°F (0°C), sitting unprotected in a closed tent is also uncomfortable. A tent offers very limited protection and insulation. Traditional greenhouses work similarly for plants; they overheat during the day if uncontrolled, and then they let all that heat out at night. The result is wild temperature swings that stress or kill plants. To compensate, greenhouse growers often blast the greenhouse with heating and cooling systems in order to grow year-round.

The reason for these inefficiencies has to do with some basic principles of design. Traditional greenhouse design focuses on maximizing light by maximizing glazing. (Glazing is a term for any light-transmitting material, like glass or clear plastic.) Traditional greenhouses are normally "100% glazed," meaning all surfaces are made of clear or translucent materials. While they are good at letting in light, glazing materials are extremely poor at retaining heat. You've experienced this first-hand if you've ever sat next to a window on a cold night—it's a chilly spot. Now imagine an entire building made out of windows. It naturally gets very cold if not heated through the winter.

Solar greenhouse designs takes a different approach. Instead of creating a fully glazed structure, it finds a balance between glazing and insulation in order to create a more thermally stable structure (one that naturally resists overheating and overcooling). Designers use glazing *strategically*, placing and angling it to maximize light while reducing the glazing area as much as possible to minimize heat loss. Furthermore, solar greenhouse design emphasizes *storing* the excess heat of the greenhouse during the day and using it for heating at night. Instead of ventilating excess heat outside, only to have to re-heat the structure at night, solar greenhouses rely on the simple greenhouse effect for heating—using the heat from the sun that is collected and trapped in the greenhouse during the day. Instead of fossil fuels, the sun provides the energy; the greenhouse collects and stores that energy, providing its own heating when it's required.

The Many Meanings of Solar

The word "solar" is an incredibly broad term—meaning relating to the sun—but it conjures up some specific images. When most people hear "solar," they picture a building with solar photovoltaic (PV) panels.

Greenhouses can include solar panels to generate renewable electricity; however, a much wiser use of the sun's energy for heating is through *passive solar design*: the practice of using solar energy for heating without relying on any electrical or mechanical devices. Specifically, it advocates carefully enhancing solar gain and minimizing heat loss in order to reduce or eliminate the dependence on fossil-fuel-based heating/cooling.

Though passive solar heating does not use electricity, in can be applied to buildings that do. Today in the building industry, a passive solar home generally refers to a house that utilizes passive solar design. These homes usually still have electrical appliances, like a refrigerator or washing machine. Similarly, solar greenhouses rely on passive solar heating, but they often have some electrical components. Many of these electrical systems transfer heat from the greenhouse to a storage medium, like the soil or water, allowing the greenhouse to take full advantage of the powerful greenhouse effect. The term *passive solar greenhouse* is often used to more explicitly describe a greenhouse that uses passive solar heating and has no electrical components at all—so it uses no electricity.

As you can see, there are some overlapping terms, so we should clarify: In this book we use the word *passive* on its own to describe systems that don't use electricity. *Active* is shorthand for systems that require electricity, like fans or pumps. For us, "solar greenhouses" are those that rely on passive solar design, and can be electrical or nonelectrical structures.

The Seven Principles of Solar Greenhouse Design

Solar greenhouses vary in almost every way—their shapes, styles, sizes, building methods, and technologies. However, there are a few unifying elements that apply to them all. To put them in a nutshell (because every book needs a nutshell), we've distilled them into these seven best practices:

1. **Orient the greenhouse toward the sun.** In the Northern Hemisphere, the majority of the glazing should face south to maximize exposure to light and solar energy.

2. **Insulate areas that don't collect a lot of light.** In the Northern Hemisphere, the north wall of the greenhouse plays a minor role in light collection. It should be insulated in order to reduce heat loss, creating a more thermally stable structure.

3. **Insulate underground.** Insulating around the perimeter of the greenhouse allows the soil underneath it to stay warmer, creating a "thermal bubble" underneath the structure that helps stabilize temperature swings.

4. **Maximize light and heat in the winter.** To grow year-round without dependence on artificial lights or heaters, it is crucial to maximize naturally occurring light and heat during the colder months. This is done by using proper glazing materials and angling the glazing for winter light collection—in general, using the glazing area strategically.

5. **Reduce light and heat in the summer.** Growing during the warmer months can create problems with overheating. Strategic shading, glazing placement and angles reduce unnecessary light and heat in the summer.

6. **Use thermal mass (or other thermal storage techniques).** Thermal mass materials are materials that store the excess heat in the greenhouse during the day and slowly radiate it at night or when needed. This evens out temperature swings, creating a more controlled environment for growing. Almost all solar greenhouses have some mechanism to store heat, broadly called thermal storage.

7. **Ensure sufficient ventilation.** Natural ventilation ensures a healthy plant environment and controls overheating.

The Case
for Solar Greenhouses

Fig. 1.1 shows the temperatures in two unheated greenhouses over a few cold, winter days in Boulder, Colorado. The first is an uninsulated greenhouse, made out of a PVC frame and polyethylene plastic. The second is an insulated solar greenhouse designed with the principles listed above.

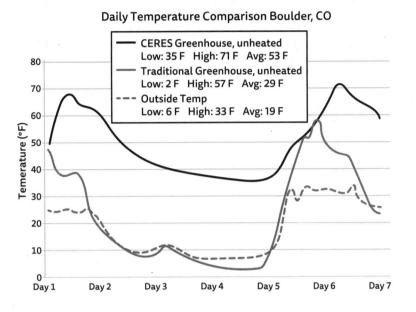

Daily Temperature Comparison Boulder, CO

- CERES Greenhouse, unheated
  Low: 35 F   High: 71 F   Avg: 53 F
- Traditional Greenhouse, unheated
  Low: 2 F   High: 57 F   Avg: 29 F
- Outside Temp
  Low: 6 F   High: 33 F   Avg: 19 F

FIGURE 1.1.

The standard greenhouse drops to a low of 2°F (−17°C); the solar green-house stays above freezing.

Solar greenhouses are often described using nebulous terms like *high-performance* or *energy-efficient*, but this is what it simply comes down to: they are able to stay much warmer year-round, and thereby grow much more than conventional greenhouses—without relying on fossil-fuel heating. They also overheat less, because they do not have excessive areas of glazing. Hence, they maintain a more stable growing environment, conducive for plants and able to grow year-round, even in harsh environments.

We've addressed the top two most common questions about solar greenhouses, now let's address a third: Do they get enough light? People often note that solar greenhouses look more like sunrooms or sheds than greenhouses. Indeed, they usually have less glazing because they work by balancing the glazed area with insulation for efficiency. However, contrary to what you might expect, they still receive roughly equivalent or even greater light levels than conventional structures. This has to do with the directional nature of sunlight and the placement of glazing, a topic discussed in Chapter 5. When light enters a solar greenhouse, rather than being transmitted through the north wall, it is reflected back inside by an insulated north wall (usually painted white).

The effectiveness and production potential of solar greenhouses has been documented in research trials—and thousands of backyards—for decades. Notably, in the early 1970s, The Brace Institute at McGill University conducted a unique side-by-side study comparing a conventional greenhouse with one built according to solar design principles. Made out of double-layer polyethylene plastic on all sides, the conventional greenhouse served as the control. The experimental solar greenhouse, called the Brace greenhouse, featured an insulated north wall, a double-layer plastic south wall and several other efficiency features. Both operated over a few seasons, and key data—temperatures, light levels and yields—were recorded. The Brace study found that light levels inside the solar greenhouse during the winter were comparable to the fully glazed structures. They were high enough to grow as much or more than conventional structures.

Here are some of their key findings:

- "The new design has yielded significant savings in energy requirements, of up to ⅓, compared to the conventional greenhouse."
- "Total weight of fruit produced in the Brace greenhouse was three times that produced in the control greenhouse."
- "Frost does not occur in the Brace greenhouse until one month after frost had destroyed the crops in the standard greenhouse."[2]

The Need for Solar Greenhouses

The greenhouses referred to in Fig. 1.1 were both residential structures; however, commercial greenhouses encounter the same problems. Typically, energy costs are the third largest expense for commercial greenhouse growers in the US (behind labor and plant materials). As of 2011, 70% to 80% of energy costs went to heating the greenhouse through cold North American winters.[3] Moreover, because of the inherent inefficiency of most greenhouses, these energy costs are vastly greater than for other types of buildings, making it challenging to grow year-round profitably. For instance, currently, the heating/cooling costs for commercial year-round greenhouses in Colorado are $3–4 per sq. ft.[4] In comparison, the heating/cooling costs for an average Colorado home are between $0.10 to $0.50 per sq. ft.[5]

As a backdrop to this situation, our agricultural system is precariously dependent on fossil fuels. For every calorie of food on your table, it took an average of *ten* calories of fossil-fuel energy to produce it. Every step of the food production chain relies on fossil fuels, from growing (pesticides and fertilizers), to processing (emulsifiers, additives, preservatives), packaging (plastic containers), and transportation. For many fruits and vegetables, shipping increases the 10:1 ratio of "energy in" to "energy out." For example, "97 calories of transport energy are needed to import one calorie of asparagus by plane from Chile [to the UK], and 66 units of energy are consumed when flying one unit of carrot energy from South Africa."[6]

Combined with volatile oil prices, finite oil supplies, and a warming planet, these statistics present a grim picture. Greenhouses are just one of many solutions that reduce the energy dependence of our food supply and re-localize food production. However, the current design of greenhouses has the potential to only shift the problem, not solve it. Though many greenhouses provide local crops, the inefficiency of the structures can undermine the effort. For example, a study conducted by Cornell University compared the total energy needed for growing tomatoes in greenhouses in New York for local markets versus growing tomatoes in fields in Florida and shipping them to New York. Taking into account production and transportation, tomatoes grown in standard greenhouses used about six times more energy than the shipped tomatoes. Though greenhouses created a local food supply, they increased the total demand for fossil fuels.[7]

Solar greenhouses hold tremendous potential as a way to reduce both food miles and fossil-fuel use, for commercial and home growers alike. The nature of solar greenhouses as warm year-round structures enables backyard gardeners to grow crops (like bananas, mangoes,

Many farmers are interested in greenhouses; what scares them the most are the heating bills.

—Steve Newman, Colorado State University, Greenhouse Extension avocados and vanilla) that are normally shipped thousands of miles across oceans. Unlike conventional greenhouses, which often struggle to stay above freezing, solar greenhouses greatly expand what we can grow, in any climate by harnessing the sun.

Endnotes

1. Dan Chiras, "Off-Grid Aquaponic Greenhouses," DVD. www.evergreen institute.org
2. T. A. Lawand, "Solar Energy Greenhouses: Operating Experiences," Brace Research Institute, Macdonald College of McGill University, July 1976.
3. Scott Sanford, "Reducing Greenhouse Energy Consumption: An Overview," University of Wisconsin-Madison College of Agricultural and Life Sciences, 2011, articles.extension.org
4. Personal communication with Steve Newman, Colorado State University.
5. US Energy Information Administration, "Household Energy Use in Colorado," data from 2009, eia.gov
6. Norman J. Church, "Why Our Food Is So Dependent on Oil," published by Powerswitch (UK), April, 1, 2005. Available at resilience.org
7. D. S. de Villiers, et al., "Energy Use and Yields in Tomato Production: Field, High Tunnel and Greenhouse Compared for the Northern Tier of the USA (Upstate New York)," Cornell University, Ithaca, NY, goo.gl/OLCBu9

CHAPTER 2

Growing Indoors:
History and Future Trends

A Brief History of Greenhouses

In your mind's eye, go back to the Roman Empire around 30 AD. Harvested food was limited to grains and a few basic vegetables when they were in season. It was around this time that a doctor of Emperor Tiberius Caesar ordered the ruler to eat a cucumber every day for good health. To provide the cucumbers year-round, growers devised a strategy of growing them underneath thin layers of a translucent stone (mica) to protect the crops. That's the first documented occurrence of a greenhouse-like structure. Though revolutionary, the invention did not catch on until about 1,400 years later. Greenhouse development has always been tied to the evolution of light-transmitting materials, and until the Industrial Revolution, the available material—glass—was a luxury.

It wasn't until the 1700s that innovations in in glass manufacturing made greenhouses possible, to an extent. Still incredibly expensive, greenhouses were only available to wealthy Europeans. Glasshouses of the Victorian era were ornate structures made out of iron frames and single pane glass. If we were to give this era another name, it could be "pomp and glass." Greenhouses were not used for serious food production. Rather, they housed exotic plants and served as status symbols for the incredibly rich, or centers of education.

FIGURE 2.1.
Conservatory
of Flowers,
San Francisco.

When greenhouses came to North America, they mimicked the European designs, despite major climatic differences between the two continents. In contrast to Northern Europe's maritime and cloudy climate, North America has harsher winters and much higher light levels. For example, the Netherlands has roughly half the heating degree days (a measure of heat requirement) than Quebec, and 40% less annual solar radiation.[1] Without much consideration given to these differences, glass greenhouses based on European design percolated across North America. The Conservatory of Flowers in San Francisco, for instance, closely resembles the famous Kew Gardens outside of London; both were built in the mid-1800s, and both were symbols of the Victorian greenhouse era.

Energy-efficient design probably would not have made a huge difference in performance for these early greenhouses because the only glazing available was single-pane glass, which sealed poorly to the frames.

In order to grow year-round in these extremely leaky and inefficient structures, early greenhouses were heated with coal furnaces or huge amounts of manure-based compost. (The compost was added to the growing beds to keep roots warm and plants alive through the winter, a method some greenhouses still use today.)

The next major development in greenhouse design came after WWII, with the advent of large-scale plastics manufacturing. The game-changer was polyethylene film, which has enabled much of our plastic-laden life: polyethylene makes shopping bags, water bottles, and a myriad of other single-use plastic products. For greenhouses, polyethylene provided a lightweight malleable covering that can be easily rolled over a thin greenhouse frame. Compared to glass, polyethylene was enormously less expensive, making greenhouses available to a huge range of growers.

As a result, greenhouse production took off, with many farmers building simple polyethylene hoop houses and greenhouses. We are still in that era today. In today's greenhouse industry, some estimate that over 90% of structures are made out of polyethylene film.[2] To accommodate for poor insulation, greenhouses are often heated with propane or natural gas and cooled with large venting systems.

While plastic greenhouses were becoming more popular, another trend emerged in the 1970s, the decade of the oil embargo and the consequent fervent interest in renewable energy. This is when the term "solar greenhouse" came into being, as some groups started looking at the energy expenditures of greenhouses and realizing that instead of fossil fuels, greenhouses could run completely off solar energy. Universities and research organizations like The Brace Institute in Montreal and The New Alchemy Institute in Cape Cod, Massachusetts, began applying passive solar design principles to greenhouses, combining insulation, strategic glazing and thermal mass. Both organizations proved that solar greenhouses could be as productive as standard designs, and were far more energy efficient. Solar greenhouses were most popular during that period, as evidenced by the slew of books on solar greenhouse design published in the 1970s and 80s (many referenced throughout this book).

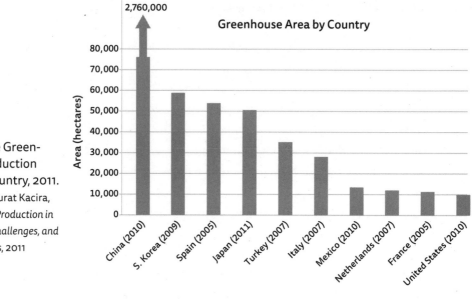

FIGURE 2.2.
Worldwide Green-
house Production
Area by Country, 2011.
Credit: Dr. Murat Kacira,
*Greenhouse Production in
US: Status, Challenges, and
Opportunities,* 2011

Greenhouses Today

Today, there are over 3 million hectares (an area roughly the size of Maryland) under greenhouse production worldwide. Though we commonly associate Northern European countries with greenhouses, by far and away the largest greenhouse-growing country is China. Strikingly, it is home to over 90% of world's greenhouse area, about 50 times more area than the next closest country, North Korea.[3]

From the global perspective, solar greenhouses are an infinitesimal part of the picture. This raises the question of why: If solar greenhouses have been documented to reduce energy costs and increase yields, why are they not more popular? One reason is that most regular greenhouses are simple, low-cost structures used primarily for season extension. They provide a little extra crop protection but are not heated, year-round structures.

Another reason is that commercial greenhouses are often incredibly large, spanning many acres. Passive solar design principles are less relevant on this scale. The energy savings of an insulated north wall, for example, become much lower when the square footage of the greenhouse is increased by a factor of one hundred. Thus, part of it has to do with the predominant system of growing monocultures at immense scales, rather than distributed growing in smaller farms.

Yet another reason is economic, or cultural, depending on how you look at it. Heated greenhouse operations rely on cheap, fossil-fuel heating. Greenhouses have been heated for many years with low-cost propane. Until recent decades, there was no major incentive to curtail energy use.

In the residential market, solar greenhouses may not have caught on because many first-time greenhouse buyers may not know the on-going costs of a traditional greenhouse. We find that many first-time greenhouse growers are surprised by how an uninsulated greenhouse performs over the first winter. Or, they may be reluctant to front the capital needed to build a more energy-efficient structure even when it would be cheaper over the long term. They may not know how effective solar greenhouses can be, or how to build one. At least in these respects, this book can help.

The Future of Controlled Environment Agriculture

The past decade has witnessed a cultural shift in how we think about growing and sourcing food. People have begun to look at our fossil-fuel-laden system of growing and distributing monocultures and realize—as farmer Joel Salatin puts it—"this ain't normal." The local food move-ment is responsible for the fact that the number of US farmers markets has more than doubled in the past decade, and organic produce is the fastest growing area of agriculture.

We are in an incredible time of innovation when it comes to growing methods to meet the demand for local food. Greenhouse production is becoming more advanced, with tools for making structures more auto-mated and controlled.

Nowhere is the desire for control more evident than in plant fac-tories, completely enclosed environments in which trays of plants are stacked vertically and grown under LED lights. Light, water, CO_2 and temperature are precisely controlled by artificial systems. These indoor farms specialize in creating perfectly controlled environments, and end up resembling factories or surgical rooms more than farms. Currently, plant factories are far too expensive to compete with traditional farming or greenhouse growing; however, growing more space-efficiently and with more automation will continue to be important trends, given the increasing pressures on arable land and water resources.

Another trend is the rise in urban food production. Spurred by the demand for fresh and local food, farms are coming to the cities—demonstrated by case studies like the GrowHaus (Chapter 18) and Growing Power (Chapter 4). These farms tend to use efficient growing methods like hydroponics and aquaponics to bring nutritious food directly to people, often those with no access to fresh or healthy food.

Another prominent trend is farmers' motivation to reduce energy and water use. Hydroponics and aquaponics—which use only one tenth as much water as conventional agriculture—are becoming larger players in controlled environment agriculture (CEA). Many other facilities turn waste, like compost, into value-added resources. The Plant in Chicago, for example, is home to several sustainable food businesses that operate

FIGURE 2.3.
Growing Power
Vertical Farm.
Credit: Growing Power

symbiotically in a repurposed 93,000 sq. ft. former meat-packing plant. The waste streams of one operation (such as a brewery's spent grain) serve as inputs to another (feed for an anaerobic digester) to create a zero-energy, zero-waste food system. Such closed-loop systems and recycling of resources are trends we hope will continue in the future of CEA.

Farms are starting to conserve space as well. The term *vertical growing* can be interpreted as growing stacks or trays of plants in single-story buildings, or multi-story greenhouses, like the five-story vertical farm proposed by Growing Power in Milwaukee Wisconsin, shown in Fig. 2.3. However it is applied, vertical growing is part of a trend of maximizing yields in small spaces. So far, skyscraper farms mostly exist on paper, but we are quickly moving in that direction, with larger and taller urban farms opening up every year.

Endnotes

1. T. A. Lawand, et al., "The Development and Testing of an Environmentally Designed Greenhouse for Colder Regions," Brace Research Institute, McGill University, 1974.
2. Jim Nau (Editor), "Ball Red Book: Crop Production." Vol 2. Ball Publishing, 2011.
3. M. Kacira, "Greenhouse Production in US: Status, Challenges, and Opportunities," Presented at CIGR 2011 conference on Sustainable Bioproduction, September 19–23, Tokyo, Japan.
4. Mission Statement, Bulletin of the New Alchemists, 1970.

Case Study: New Alchemy Institute
Forty Years of Growing: The Evolution of "The Ark"

**1,800 sq. ft. residential and R&D greenhouse
Cape Cod, Massachusetts**

In 1969, a group of scientists formed The New Alchemy Institute to explore sustainable ways of living and producing food. Housed on a former dairy farm in Cape Cod, Massachusetts, the group pioneered research into aquaculture, organic growing, composting, permaculture and bio shelters, following the belief that "ecological and social transformations must take place at the lowest functional levels of society if human-kind is to direct its course towards a greener, saner world."[4] While these are common house-hold terms today, they were fringe concepts in the 1970s. The New Alchemists conducted some of the first scientific studies into more sustainable ways of living and growing food.

One of their primary goals was to maximize food production in small spaces without relying on pesticides or fossil fuels. In 1971, their first greenhouse, the "Ark" began as a simple plastic dome over a wading pool filled with tilapia. Over the next several years, it evolved to resemble the structure it is today—an 1,800 sq. ft. solar greenhouse with integrated solar panels, a solar hot water system, aquaculture and a year-round permaculture garden. The Ark was one of the first structures to incorporate elements of pas-sive solar design into a greenhouse, including a well-insulated north wall, south-facing glazing, and a great deal of thermal mass in the form of several large fish ponds.

After the New Alchemists disbanded in the 1990s, two of the original members, Earle Barnhart and Hilde Maingay, purchased the property and started the nonprofit The Green Center to continue the mission of the New Alchemists, where they still research more sustainable ways of living. Forty years after it was built, life on the site still revolves around the Ark. In 2000, Barnhart and Maingay built a super-efficient home onto the greenhouse, like a normal greenhouse addition in reverse. In addi-tion to providing most of their food, the green-house houses a solar PV system, which powers the greenhouse and their home, and a solar hot water system to provide hot water and much of their home heating.

FIGURE 2.4. The Green Center (Formerly New Alchemy Institute).

CHAPTER 3

Planning for the Greenhouse

Don't fight the forces; use them.
—Buckminster Fuller, *Shelter*, 1932

Season Extension Options

Chapter 2 laid out the very big picture. Now, let's zoom in to evaluate the range of possibilities for extending the growing season. This book focuses on solar greenhouses, which are structures that feature an insulated north wall, double layers of glazing, underground frost protection, and various methods of thermal storage. Solar greenhouses are designed to stay much warmer than outdoor temperatures, creating lush, abundant, year-round growing environments—enabling growers to grow and experiment with many crops normally ungrowable in their climates. However, they are not the only way to extend the season. Simpler methods of crop protection also have a place in the range of solutions, summarized in Fig. 3.1. Understanding these options provides context for where solar greenhouses fit in, and whether they are the best fit for you.

Hoop houses, row covers and cold frames all provide a single layer of crop protection and usually used as season-extenders. They provide some frost protection for crops, creating an indoor environment a few degrees warmer than outside. Studies show inside temperatures under a

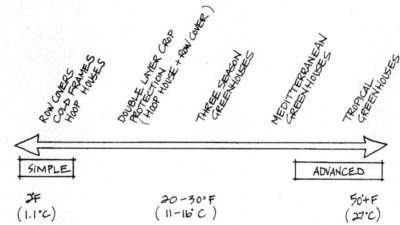

FIGURE 3.1.
Season
Extension
Options.

single layer of protection are on average 2°F–4°F warmer than the minimum outdoor temperature.

The next step is to add a second covering using row covers in addition to the outer structure, as shown in Fig. 3.2. Eliot Coleman is well known for using this technique to grow cold-hardy greens year-round on his farm in Maine. The added layer protects crops enough to sustain hardy greens even when the outdoor temperatures drop to −8°F (−22°C). One of the keys to Coleman's successful winter farm is growing crops that tolerate frost and low-light levels. This supports Coleman's basic philosophy of simplicity and limited intervention. For other growers, the goal is produce more of the food we consume—which goes far beyond cold-hardy greens.

To do this, we turn to greenhouses that can withstand much greater outdoor temperature extremes and maintain a stable year-round growing environment. This is where solar greenhouse design comes in. By trapping heat and retaining it overnight (using insulation and thermal storage), a greenhouse can maintain an environment above freezing for all or most of the year, expanding both the growing season and the variety of crops we can grow.

Within the category of solar greenhouses, there is also range of options based on the design and growing goals. One way to frame the

FIGURE 3.2.
Row Covers Inside
a Hoop House.
Credit: Mehaffey Farm

options is by their minimum indoor temperature, just as outdoor growing zones are categorized. Some solar greenhouses are three-season
structures that freeze over in the winter. This is a common strategy
among growers in climates with very harsh and low-light winters (for
examples, see the case studies from Canada and the northeastern US).
Then there are greenhouses that get *close* to frost in the winter. These are
often called *Mediterranean greenhouses*, reflective of the milder winters
of that region. These greenhouses enable many more things to grow
year-round: cold-tolerant vegetables and perennials, and trees like figs,
olives and varieties of citrus.

Finally, there are the hothouses, also called *tropical greenhouses*.
These have minimum temperatures of 40°F–50°F (4°C–10°C), which
permits growth of heat-loving crops like tomatoes, peppers and egg-
plant as well as perennials like bananas, guava and citrus. This type of
greenhouse requires more insulation, multiple layers of glazing, and significant thermal storage to create an indoor environment that is several
growing zones above the outdoor growing zone, as shown in Fig. 3.3.

Light is usually the limiting factor for growth, making good growth of many full-sun crops difficult in northern areas. Which season extension options work for you depends on your climate, your goals for the growing environment, and your resources.

Understanding Your Climate

One of the reasons traditional greenhouses perform poorly in most climates is that they use a "one-size-fits-all" approach. The same plastic box will operate very differently in Maine than it will in Texas. Solar greenhouse design applies a different mindset: by tailoring the structure to the local environment, one can work with the elements, rather than against them.

To design a structure that works cohesively with the outdoor environment, you must know a bit about your local climate, both its challenges and resources. From a greenhouse design perspective, the two most important variables that make up the local climate are *light* and *temperature*.

There are a variety of ways to describe and measure each. Regional temperatures can be characterized by their averages or their minimums, or by other metrics like Heating Degree Days (which reflects the energy requirement to heat a building). For the purposes of this book, we find that the USDA Growing Zone map, shown in Fig. 3.3, is most helpful as a description of temperature zones. It categorizes climates by their minimum temperature. You can find versions online that show much more detail by state.

Like temperature, there are many ways to measure light and many factors that contribute to the light levels at a given location: percentage of possible sunshine or cloudiness; the intensity of sunlight based on latitude and elevation; day length, etc. We delve into this subject much more in Chapter 5. In short, the simplest metric greenhouse growers use is the daily light integral (DLI), which measures light intensity over a period of time. The DLI integrates all these factors into a simple number indicating the total light levels available to plants over a 24-hour period. You can see DLI numbers for the US in the first map in the color section.

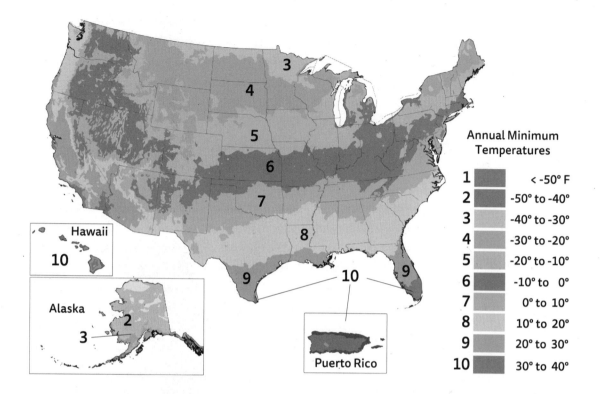

FIGURE 3.3.
USDA Grow Zone Map.
Credit: US Department of
Agriculture

Defining Your Goals

The outdoor climate is only one element to consider. On top of this, you also need to decide what kind of *indoor* environment you want to create. Some basic questions can help elucidate. You don't have to have all the answers right away; the important thing is to have these questions in the back of your mind as you move through the design process so you can develop clear goals to inform your greenhouse plan.

- **What do you want to grow and when?** Your goals for your indoor environment will inform all manner of decision in the design process. Do you want to grow tomatoes year-round? Then you need a greenhouse that stays above 50°F (10°C). Do you just want to grow greens in the winter? If so, a three-season greenhouse that avoids freezing can suffice. There is a huge difference between those two growing environments, and thus the structures/systems required to create them.

Often, growers look for groups of crops that have similar temperature requirements, such as warm-season annuals, cool-season annuals, and/or different perennials that have similar needs, and then they design the greenhouse to meet the minimum temperatures required by that group. A list of common greenhouse crops and their temperature ranges is given in Appendix 1.

- **Why do you want a greenhouse?** Are you growing as a hobby or commercially? Commercial greenhouses typically require narrower temperature ranges, which necessitates more advanced climate control systems. In a residential setting, a freeze or loss of crop is not catastrophic, and more variation is tolerable.

- **Have your considered other uses for your greenhouse?** Clearly, you want to grow plants, but could it also be an area for sitting/relaxing, education or storage? Might you integrate animals into your greenhouse? We recommend you start by sketching out a floorplan early on in the process, and allow this to evolve as you refine your design.

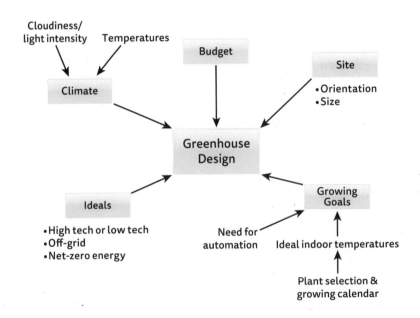

FIGURE. 3.4.

- **What is your time commitment?** Do you want to be able to leave the greenhouse and go on vacation, or can you check on it daily? This will determine the need for automated systems, such as vent openers or fans. On a broader level, it influences how big the greenhouse should be. Think of how much additional garden space you can manage based on your schedule throughout the year. To decide on

Case Study: Verge Permaculture
Growing Year-round in the Canadian Rockies

200 sq. ft. residential greenhouse
Verge Permaculture
Calgary, Canada

Formerly pipeline engineers, Rob and Michelle Avis founded Verge Permaculture in 2010 after realizing they wanted to "change the status quo, not support it." They converted their Calgary home into a resilient homestead based on permaculture design and started developing educational tools to help others do the same. When they decided to add a greenhouse to their property, they knew they were up for a challenge. Average temperatures in the winter are 10°F–26°F (–12°C to –3°C) and can drop to –31°F (–35°C) on cold nights.

To accommodate, Rob and Michelle built a super-insulated greenhouse out of SIPs (structurally insulated panels) and double-wall polycarbonate. After operating it for a couple of years, they decided that trying to get full production in the winter wasn't practical in their climate. The primary limitation was not a lack of heat, but light. The slower growth in the winter didn't justify operating the greenhouse.

Rob has a very practical take on year-round growing in a climate like Calgary's. "If you plan to grow in the winter, returns on production diminish rapidly as labour and other inputs increase dramatically. My goal to only extend the regular growing season saves time and energy by giving the greenhouse a winter's nap."

The Avises start planting it in early spring, and they have ripe tomatoes by July (about the time when it's just possible to start planting outside in Calgary). They occasionally use a rocket mass heater for heating later in the season. They close the greenhouse in November, let the soil freeze out any pests, and concentrate on their permaculture consulting business over the winter.

You can read more about the Avis's greenhouse and download their ebook on passive solar greenhouse design at vergepermaculture.ca. Rob and Michele Avis's greenhouse is shown in the color photo section (titled "Verge Permaculture").

a greenhouse size, we recommend residential growers consider their outdoor gardens as a guide. Think of your past garden plots, how much time they required and how much food they provided. Do you want something larger or smaller? Commercial growers naturally have many more factors to consider—production numbers, labor requirements and expenses—all of which should be detailed on a business plan.

- **What is your budget?** The primary factors determining greenhouse cost are usually the labor and level of technology. If doing the work yourself, collecting recycled materials, and using simple systems, greenhouses can be built on a shoestring budget. If hiring out the work and using more advanced systems, costs naturally rise. As the greenhouse gets larger, the cost *per square foot* decreases.

All of these factors come together and influence each other in a web of variables that makes every solar greenhouse unique, outlined in Fig 3.4.

Further Reading

Coleman, Eliot. *The Winter Harvest Handbook*. Chelsea Green Publishing, 2009.
Jabbour, Niki. *The Year-Round Vegetable Gardener*. Storey Publishing, 2011.
USDA online interactive map for Plant Hardiness Zones, planthardiness.ars
.usda.gov

DESIGNING AND BUILDING
A SOLAR GREENHOUSE

CHAPTER 4

Siting and Orientation

Orient the Greenhouse
Toward the Sun

This is where solar greenhouse design begins: the sun. A major tenet behind all principles in this book is managing solar energy—heat and light—to create a naturally controlled year-round growing environment. To maximize light year-round, and particularly in the winter, the majority of the greenhouse glazing (light-transmitting materials) should face the direction of the sun: the south if in the Northern Hemisphere.[1]

Fact: On average, on a clear sunny day in North America, 85%–90% of solar radiation comes from the south; less than 15% is diffuse light coming from other directions.

This fact provides the reasoning behind two principal features of solar greenhouses: south-facing glazing and an insulated north wall. By using glazing on the south side of the structure, you maximize the effectiveness of the glazing area.

In contrast, surfaces facing north play a minor role in light collection during the day. If they are glazed, there is little benefit from a growing or heating perspective. Plus, the sacrifice of doing so is huge—it greatly reduces the heat loss of the structure, causing the characteristic freezing cold temperatures of traditional greenhouses.

Q
&
A

Why are many commercial greenhouses oriented facing east/west?

Large commercial greenhouses are often gutter-connected structures, which involve several units connected by ridges. If a gutter-connected greenhouse faces south, the thick ridge beams in between structures create shadows on the floor. The shadows move very little throughout the day, creating slower plant growth in that small area. Thus, even though it is a sacrifice for energy efficiency, the commercial industry often recommends facing the long axis of the greenhouse east/west to get a uniform crop in a tightly competitive industry. But for potential solar greenhouses builders, this recommendation has little relevance; it shouldn't be followed for energy-efficient structures.

Choosing Dimensions

In order to increase solar gain in the greenhouse, your goal is to maximize the area of south-facing glazing—while maintaining reasonable dimensions that fit your site.

You may read that solar greenhouses should be twice as long as they are wide. This refers to the *aspect ratio*: the ratio of length to width, as demonstrated in Fig. 4.1. While not a hard-and-fast rule, the reasoning behind it is sound: by elongating the structure, you can increase the area of south-facing glazing (the heat collector) relative to the footprint. For this reason, solar greenhouses usually have long rectangular shapes, with the long, glazed side facing south. That means the *ridge* of the greenhouse runs along the east-west axis, a common way to describe a greenhouse's orientation.

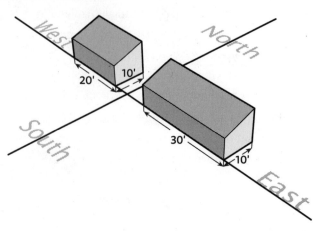

FIGURE 4.1.
Aspect Ratios
and Orientation.

Figure 4.1 shows aspect ratios of 2:1 and 3:1, length:width. We recommend something in this range, depending on what your site allows. Again, keep in mind practical considerations like your site, possible obstructions, and maneuverability inside. Larger, commercial greenhouses often have to be wider in order to accommodate a larger footprint and maintain an easily manageable space.

When it comes to the height of the greenhouse, consider local building codes, which often stipulate a maximum height for buildings. Residential greenhouses are typically in the range of 8'–11' at their peak. Commercial greenhouses with larger footprints necessitate taller walls, as do larger structures growing tall crops like fruit trees. (Some growers go as high as 17' to accommodate trees. This can stick out in a backyard, though. Our residential greenhouses are typically 9'–10' on the north wall.)

Climate Considerations

When choosing an orientation, consider your regional climate conditions and local weather patterns. For example, many areas of the East Coast of the US have clear mornings that turn to cloudy afternoons—a reason to favor a southeast orientation over directly south. Other coastal areas have cloudy mornings that burn off into clear afternoons, so tilting the greenhouse more toward the afternoon light would be best, as long as your area isn't too hot.

What If the Greenhouse Can't Face Directly South?

Not all locations have an open southern orientation, clear of obstructions. Often, a tree or a structure will shade part of the greenhouse, forcing you to shift the greenhouse away from south. The precise effect varies by location, but we generally find that shifting the greenhouse up to 45 degrees away from south—that is, southeast or southwest—does not have significant impact on total light collection. Across many different locations, energy models of greenhouses show that a shift of up to about 45 degrees in either direction allows greenhouses to still receive over 90% of the light they would get with a south orientation. However, light starts to attenuate if you shift further east or west. In Boulder, Colorado, for example, a greenhouse oriented 67.5 degrees toward the west (so facing predominantly west) only gets 70% of the annual light compared to a due-south orientation. Orienting the greenhouse any further than this dramatically reduces light levels, particularly in the winter. (The exact variation depends on your latitude and climate.) Thus, we recommend anything within 45 degrees of due south, with a preference toward southeast orientations.

Orienting the greenhouse more southeast allows for more light in the morning, which has a number of benefits. Morning light is considered softer or less intense than afternoon light because it lacks the intense heat characteristic of the afternoon sun. Moreover, in the morning the greenhouse is cool, having just gone through the coldest part of the night (just before dawn). Morning light allows the greenhouse to warm up faster, reducing stress on plants.

In contrast, facing the greenhouse southwest brings in much more heat in the afternoon, after the greenhouse has already been heating up all day. It risks overheating the greenhouse or scorching plants in front of west windows. For these reasons, most growers prefer a southeastern orientation to take advantage of the light and heat in the morning, particularly those located in warm climates. (Note that these factors also inform door/window placement. We rarely place doors on the east side of the greenhouse to allow for a larger window area there. Rather, doors can be conveniently placed on the west or north walls.)

Tools for Orienting the Greenhouse

A *compass* is needed to identify due south at your site. This being 2016, there are free apps for smart phones that perform this task.

An extra step for precise growers is to use a *magnetic field calculator* like the one available from NOAA (ngdc.noaa.gov/geomag-web) to find the variation between true south and what a compass will read. Called *magnetic declination*, there is a slight variation between the two due to the earth's magnetic field—though it's only a few degrees.

Siting the Greenhouse

The goal for siting the greenhouse is the same as for orientation: ensuring good light year-round, and maximizing light during the colder months. Light access at a potential greenhouse site is determined by macro factors (the path of the sun at your latitude) and site-specific factors, like shading from trees.

The Solar Path

In order to evaluate a site for a greenhouse, you need to know a bit about where the sun will be when. The sun's path through the sky is constantly changing over the course of the year. Fortunately, though, it is predictable. The sun's position is determined by two coordinates: the elevation and the azimuth.

- **Angle of elevation (or altitude)** describes how high the sun is in the sky from the ground plane. It changes from low in the winter to high in the summer, as shown in Fig. 4.2. This seasonal variation becomes more pronounced the further you are from the equator. We'll reference this angle extensively throughout the book, so it is helpful to jot down some key elevation angles for your location. Some quick rules of thumb: At solar noon on the equinoxes (spring and fall), the

FIGURE 4.2. Solar Altitude Angle and Solar Azimuth Angle.

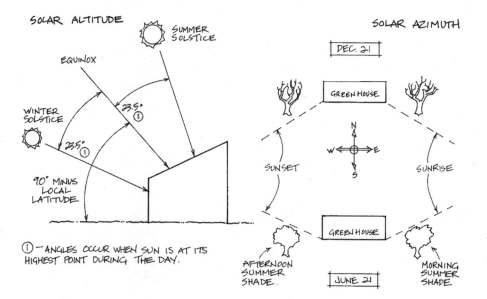

angle of the sun is equal to 90 minus the local latitude. For example, Denver is at latitude of 40 degrees. The solar altitude at noon on the equinoxes is 50 degrees (90 − 40). At the winter solstice (December 21), the solar elevation is 23.5 degrees less than the equinox (= 26.5 degrees for Denver). On the summer solstice (June 21) it will be 23.5 degrees greater than on the equinox (= 73.5 degrees for Denver). Online tools are also available (see below).

- **Azimuthal angle** is a fancy term for the angle of the sun on the horizon, relative to due north. Importantly, it indicates the position of the sunrise and sunset each day; so it changes from a very wide arc in the summer (resulting in long days) to a narrow arc in the winter (creating short winter days) as shown in Fig. 4.2.

Resources for Determining the Solar Path

The website SunCalc.net gives a visual representation of the solar elevation and azimuth overlaid on a Google map; however, it only shows one date at a time. You'll have to scroll through to get a full sense of the year. Other organizations, like the University of Oregon's Solar Radiation Monitoring Laboratory (solardat.uoregon.edu), offer software tools to create sun path diagrams. There are also apps, like Sun Surveyor, Sun Seeker and Helios Sun Position Calculator (available for iPhones and Androids), that have a lot of perks that most websites don't have, but that cost $10–$30. We have not found a single outstanding visual tool that shows the solar path; you may have to use multiple tools to get a full understanding.

Together, these two angles create the solar path—the path the sun will take every day over the course of the year. A plot of that path is called a solar path diagram, which you can look up for your location using several online tools, described below. Whatever method you choose, it's essential to have a basic understanding of how the sun's path changes, from a narrow arc and low elevation in the winter, to a wide arc and high elevation in the summer. This movement forms the basis for many decisions in the greenhouse design process.

Evaluating Shadows

Now that you have a grasp of how the sun will move at your location, you can get more site specific. Obstructions like neighboring buildings and trees can block critical winter light.

There are a few ways to evaluate shadows at your site. On the very simple end, observing shadows can suffice for residential growers. Many backyard growers already have an intuitive understanding of light and shadows at their site. Be sure to observe these during the coldest months, when light is most critical.

A second step is to create a bird's eye view sketch of the site, plotting the location of obstructions relative to the greenhouse site, as shown in Fig 4.2. Then, add the angles of the sunrise and sunsets for some key dates, such as the winter and summer solstice. This way, you can see whether the obstruction is in the "view window" of the greenhouse during the times of the year that light is needed. If it is, you can go on to calculate the obstruction's length and impact on the greenhouse at different times of year.

There are a couple ways to determine whether, or when, an obstruction will shade your greenhouse site. The math-centric approach is to use hand measurements and calculations. The basic process is to go out to your site, determine the height of nearby obstructions using an elevation gauge, and then use trigonometry to determine the length of the shadow at different times of year based on the elevation angle of the sun. Online calculators and guides can also help.

Another strategy is to measure the available light at the site, rather than shadows. This comes with the challenge that light levels change dramatically daily and seasonally, so it is hard to get a full picture. Light meters, such as the LightScout DLI 100 shown in Figure 4.4, record light levels over the course of a day. To get a full sense of light conditions, you will need to record many days. These are especially useful for comparing two different sites, as long as readings are taken on days with similar conditions. Simply move the meter to different sites to see which has greater light, or use two meters to compare two potential sites simultaneously. More on light meters in "Tools for Siting the Greenhouse."

While the goal is to reduce shading on the greenhouse in the winter, summer shadows can actually benefit the greenhouse. Depending on your climate, it's likely that there will be plentiful light for growth after the last frost date. In the summer months, light and heat in a greenhouse can get excessive and overheat the greenhouse. At these times of year, some shading is often beneficial. The best shading obstructions are deciduous trees that drop their leaves in the winter, providing light when you need it, but shading when you don't. A tree on the west side of the greenhouse, as shown in Fig. 4.4, is a natural benefactor for the greenhouse.

Other Siting Factors

Access

A good greenhouse site is also easily accessible. The more you are in the greenhouse, the easier it is to spot pest/disease problems, ensure proper watering, keep an eye on plants, and enjoy it. Thus, favor spots that give easy access, and consider creating paths or covered walkways to encourage access.

Likewise, if you are planning on using electricity in the greenhouse

Tools for Siting the Greenhouse

Light meters range from simple to advanced. Precise meters are called "quantum meters" and usually cost over $300. They measure light intensity—light at a single point in time—so you will need to take many readings to get a full idea of light levels over a period of time, or combine this with a data logger to record many readings over a period of time.

Spectrum Technologies makes a basic light meter we find very useful—at a more reasonable price for residential growers (currently $59). Called the LightScout DLI meter, you simply stick it in the ground at the greenhouse site and come back in 24 hours to get a reading of light levels in a metric called the daily light integral, or DLI (discussed in Chapter 5). The readings are very basic—the meter only gives a few ranges—but using one can be a good first step in getting an idea of light levels.

The 3D modeling program SketchUp can be used to evaluate shadows—and much more. SketchUp allows you to make a 3D rendering of the greenhouse and surrounding objects. It then gives you the option to "geolocate" the model—the program will input the solar data for your location. Once the model is created, the "shadowing tool" allows you to instantly evaluate shadows over the course of the year simply by moving a cursor. Unlike the other methods, you can see where shadows will actually fall on your greenhouse at what times of year.

Moreover, once you have a SketchUp model created, it can also be extremely useful for the planning process. If the model is detailed enough,

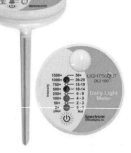

FIGURE 4.3.
Light Scout DLI
100 Meter. Credit:
Spectrum Technologies

and/or a conventional water supply (both are discussed in later chapters), consider how easy it is to access to the nearest electrical panel and water hookup.

Hillsides and Sloped Sites

South-facing slopes can be excellent sites for greenhouses. As we'll discuss more in Chapter 11, building underground increases the energy efficiency of the structure. However, it requires more time and effort for site preparation. The greenhouse has to stand up to the downward pressure of soil, necessitating retaining walls or more robust foundations. Though slightly more costly, we normally encourage building into a south-facing slope if it's available.

you can use it to show engineers, contractors and builders the plan of your greenhouse. SketchUp is available as a free download, and, contrary to first impressions, it is relatively easy to learn (no modeling or CAD experience required) by watching tutorial videos.

Finally, an accurate but costly tool is the Solar Path Finder, a simple device commonly used by solar panel installers. It's a glass dome on a tripod that you can easily move around to evaluate different potential sites. At each site, the dome shows the reflections of objects, indicating shadows. While accurate and simple, the device currently costs about $300, which may not be worth it for a single use to site a residential greenhouse, but it can be used for other projects like siting a solar hot water system.

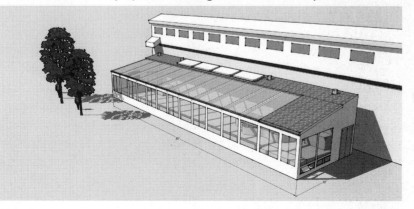

FIGURE 4.4. In addition to helping with greenhouse design, SketchUp models can show shading on the greenhouse at different times of year.

Case Study: Growing Power
An Urban, Community-Centered Farm

2.5-Acre Farm and Greenhouses
Milwaukee, Wisconsin

On a busy four-lane road on the outskirts of Milwaukee, a couple blocks from a McDonalds and Popeye's Kitchen, stands a nationally recognized and cutting edge urban farm. On this unexpected site, Growing Power has become a symbol of what's possible in urban, community-centered farming.

Founded Will Allen in 1993, Growing Power's current operations extend to many areas and locations. They include farm sites in Chicago, municipal composting partnerships, school gardens, and fundraising for a proposed five-story vertical farm.

Though multi-faceted, the Milwaukee farm centers around several year-round aquaponic greenhouses. Here, Allen has set up an operation that concentrates on space-efficient, sustainable growing practices. Several gutter-connected greenhouses grow micro-greens in stacked trays to maximize year-round production. The greens take just a few weeks from seeding to sale, providing a valuable year-round revenue stream.

The greenhouses also house a large-scale aquaponic system. Growing Power raises 5,000 fish—mostly yellow perch—annually, which it sells to area restaurants. The solar hot water system heats the aquaponic fish tanks, which in turn provide heating for the greenhouses. Natural gas heaters supplement heating. The facility also houses an 11 kW solar PV system, which provides 20–30% of the operation's electricity.

Growing Power doesn't just grow food, though; it "grows soil," as Will Allen says. Every year, the farm processes 43 million pounds of food waste, wood chips and spent grain from local breweries. Weekly truckloads deliver thousands of pounds of organic waste to the site, where it begins decomposing in massive mounds. After several months, the compost is moved to vermicomposting beds, where millions of worms turn it into nutrient-rich organic fertilizer. In the fall, volunteers also pile compost around several hoop houses. The compost heats the structures through the natural process of decomposition, allowing them to grow year-round through the harsh Wisconsin winters. After the compost is turned into high-quality soil, it's sold it to local gardeners, generating revenue as it completes the cycle from waste to valuable fertilizer.

Each of these operations could be a case study in itself. In our view, it's how they all work together that makes Growing Power such an inspiring operation. Food waste is diverted from a landfill and turned into rich compost, which is then used for growing more food. Fish and plants thrive in a symbiotic cycle in the greenhouses. Humming in the background is a diverse mix of people who tend, volunteer and learn from the site. Even with all its programs and growing methods, Growing Power's marketing literature focuses on the people it serves—the hundreds of community members involved its training and volunteer programs and thousands more Milwaukee residents who have easier access to healthy, local food.

More information is available at growingpower.org.

Will Allen, founder of Growing Power.

Permits

Whether you need a permit for the greenhouse depends on where you live, the size of greenhouse, and the joys of your local building regulations. Most counties don't require permits for freestanding structures under a certain size, usually in the 120–200 sq. ft. range. These are called "detached accessory structures," and greenhouses often pass as such. Permits are typically required for attached greenhouses (they are considered an extension of the home) and commercial greenhouses. Growers living within city limits will have greater restrictions. In all cases, it helps that solar greenhouses use standard construction methods. Like a shed, they can be engineered for any wind or snow load and can be customized to match the look of a home, which appeases building departments and home owners associations.

Takeaways

- The majority of the glazing in the greenhouse should face the direction of the sun (south for locations in the Northern Hemisphere). The north, and some of the east and west, should be insulated.
- Use long rectangular footprints to maximize the area of south-facing glazing.
- If a southern orientation is not possible, orient the greenhouse southeast, or if necessary slightly southwest. Up to about 45 degrees away from south will still yield sufficient light levels.
- Choose a site that has excellent access to winter light. If shading occurs in the summer, that is acceptable—and likely beneficial.

If meeting all these conditions doesn't sound possible, don't worry. Greenhouses are flexible environments. Even with less-than-ideal conditions, you can still produce abundant food year-round by adjusting how or what you grow.

Endnotes

1. For simplicity, we'll assume a location in the Northern Hemisphere for the rest of this book.

CHAPTER 5

Controlling Light and Heat Gain: Glazing

Solar greenhouse design depends on the balance of glazing materials (light-transmitting materials) and insulation. Historically, greenhouses have given primacy to getting the maximum amount of light possible while sacrificing energy efficiency. Smart greenhouse design relies on using glazing more carefully. By maximizing the light coming through the glazing area, you limit the amount of glazing necessary, allowing for more insulated wall area, and a more thermally stable structure overall. The purpose of glazing is to transmit solar radiation (both for heat and growth), so the first part of this chapter provides some context on the electromagnetic radiation we call light.

How Plants See the World

A thing to remember about sunlight: we can't see all of it. We see a small slice of the light spectrum, called the visible spectrum. Similarly, plants also use only a small portion of the spectrum for growth. This is called photosynthetically active radiation (PAR) light. Conveniently, it covers about the same range that humans can see, roughly 350 to 700 nanometers in wavelength. While the range is the about the same, plants and people are sensitive to different parts of the spectrum, as shown in Fig. 5.1. Our eyes are most sensitive to colors in the green range (with a peak at 550 nm). Plants are most productive when exposed to the red and the blue portions of the spectrum (so, the part we are least sensitive to).

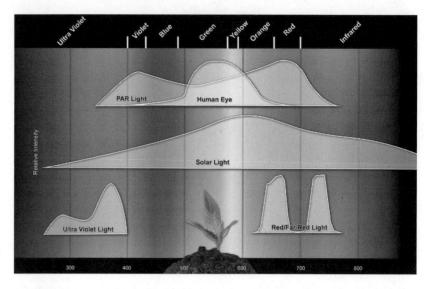

FIGURE 5.1: Photo-synthetically Active Radiation vs. Visible Light. Credit: Spectrum Technologies

How does this relate to building a greenhouse? First, it means you can't accurately evaluate a glazing material just by looking at it. Our eyes are not great sensors for the light that plants use, PAR light. Some new residential windows, for instance, contain tints or coatings that transmit visible light but block most of the ultraviolet (UV) and infrared (IR) portions of the spectrum; in doing so they can block portions of the PAR spectrum. Secondly, it means that the units of measurement that apply to visible light (foot-candles and lumens) are not accurate tools for measuring and evaluating light levels in a greenhouse.

Do plants use infrared (IR) or ultraviolet (UV) light?

For the most part, no, plants don't use IR or UV light directly for growth. However, these wavelengths have other effects on the plants and the greenhouse. IR light is useful for heating the greenhouse and thermal mass in the winter. Far UV light (UV-B and UV-C) is actually harmful for plants, causing DNA damage and mutations, just like it does to our skin. For this reason, and because it degrades plastics, many glazing materials intentionally block UV light. UV-A light (close to the blue range) can bring out some unique characteristics in plants, like deep purple hues, but it is not essential for growth. In general, glazing materials that have UV-blocking coatings are helpful, as long as they transmit the full spectrum of PAR light.

Measuring Light

There are myriad ways to measure and describe light. Here we'll just graze the surface of this intensive subject in order to make informed decisions in the greenhouse planning process. It's helpful to have a basic understanding of lighting units so that when you see one (say, on a glazing spec sheet or light bulb package) you can put it into context.

First, it is important to note that light is like rainfall; it changes in both intensity and duration. Measurements of both are needed to get a full picture of the total light available at a site, or to describe how much light plants need. The intensity is often called light *quantity*; duration is called the *photoperiod*. Light *quality* is a term used to describe the spectrum of wavelengths, or color, which can be affected by a glazing material.

All three of these factors determine the productivity of plants in the greenhouse, but most lighting terms describe quantity, the intensity at a single point in time. Furthermore, measurement units vary by the spectrum they measure. One way to frame the many terms is by what they are used for, either people, plants or systems.

Light for People

Lumens, lux and *foot-candles* all measure the intensity of the visible spectrum. They are calibrated to measure the wavelengths we are most sensitive to. Thus, they are limited tools for measuring light for plants; they don't measure the PAR spectrum, and they only measure light at a single point in time. Ft-candles are useful for comparing light conditions, as shown in Fig. 5.2, and are commonly used to rate grow lights.

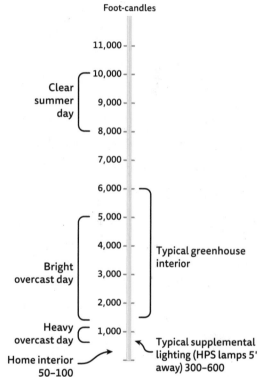

FIGURE 5.2. Comparison of Light Intensities in Ft-Candles. Though not a good measure of the total light available at a site, ft-candles are useful for comparing different conditions, such as a clear versus overcast day. Some say that plants need at least 2,000 ft-candles for growth, but this is an overly simplistic metric because it doesn't take into account the length of light exposure.

Light for Systems

Sunlight can also be measured in *watts,* a measure of power. Watts can measure the full spectrum, including the IR and UV wavelengths, going far beyond the visible or PAR spectrum. Watts per square meter is commonly used to rate power systems, like solar photovoltaic panels. For instance, on a bright day the total amount of solar radiation hitting the ground is about 1,000 watts per square meter. If you have 10 square meters of solar panels on your roof, and the panels are 15% efficient, you could estimate that the panels would produce 1,500 watts, or 1.5 kilowatts at that time (10 square meters × 1,000 watts × 0.15 = 1,500 watts). Watts can be converted to *Btus/hour,* another common metric for rating the heat gain of buildings (1 watt = 3.4 Btu/hour).

Light for Plants

The most important term for greenhouse growers is the *daily light integral* (DLI). DLI measures the PAR spectrum (using a unit called a micromole). More importantly, the DLI measures the *total amount* of light over the course of a day. Conceptually, it's is like a rain gauge; it combines both intensity and duration (a 24-hour period) to get a single cumulative metric for light. That makes it very useful for comparing light levels and describing how much light plants need, making it the most common metric in the horticultural industry, and the most helpful one for any grower to use.

The first map in the color section shows the average DLI values by month across the US. They vary from 0–60 depending on location and time of year. By looking at the DLI numbers for your location, you can see the huge fluctuations in light levels. On average in the US, the winter months have only 10%–20% of the light as the summer. Higher latitudes will experience greater variations.

This presents the basis for one of the main principles of solar greenhouse design: maximize light and heat in the winter. If a plant that evolved to grow during the long days of summer is in an environment with only one tenth the light levels, there will obviously be repercussions. Transmitting enough light and heat in the low-light months is a main goal of glazing in a year-round greenhouse.

How Much Light Do Plants Need?

This is a common question, but a hard one to answer, because it all depends on your expectations. If you were to ask "how much money do I need to live?" you would first need to know whether you want to live in a camper van or a castle. Similarly, what you want to grow, and how robust you expect this growth to be, determines how much light you need.

Though it's a subjective question, there are a few studies that give some context to the "how much light" question (see Further Reading).[1] The drawback is that these studies are targeted to the commercial greenhouse industry, which has a different set of expectations. Many commercial growers consider a DLI of 12 to be the minimum threshold for good growth of most crops. For high-light crops like tomatoes and peppers, commercial growers often shoot for a DLI of 20. (Many fruiting crops like tomatoes survive given warm temperatures, but do not set fruit in low-light.) However, home growers, and even many commercial growers, can grow year-round with much less light than this, going down to the low single digits of DLI levels. Because light requirements are subjective, based on your expectations for growth, we recommend experimenting the first year by planting many different crops, then observing and recording what you find works best in your particular conditions.

Growing in Low Light: Strategies for Winter Growing

Many growers in high northern latitudes discover that heat is not the limiting factor when it comes to year-round growing; it's the light. Dealing with low light levels is a fact of life for most year-round gardeners. Beyond designing the greenhouse to maximize light, there are two strategies to compensate: adjusting how and what you grow; and/or using artificial lighting.

The first is the more natural—and energy-efficient—method. By timing your growing pattern you can create robust year-round growth. Starting plants in the late summer/early fall alows them to easily "ride out" the low-light months. (We'll return to growing calendars in Chapter 17, "Creating the Greenhouse Environment.") In addition, cool weather crops tolerate low light better than warm weather crops (also called full-sun plants). Fruiting plants can survive in the right temperatures, but they often don't set fruit in low light.

Other strategies for growing in low light:

- **Don't let the greenhouse overheat in the winter.** It may seem only natural that a greenhouse will be cooler in the winter, but in sunny areas, greenhouses can overheat just as easily in the winter if they are completely closed. Prolonged high temperatures (over 80°F [27°C]) induce the plant to grow quickly, even though there isn't sufficient energy for photosynthesis, producing the "spindly" effect. According to the *Ball Redbook: Crop Production*, "cool temperatures [during low-light periods] allow the leaves and flowers to develop slowly, which allows the plant more time to accumulate energy from sunlight to produce healthy leaves and flowers." Keeping greenhouses cool under low-light conditions is one of the reasons why Northern European countries are able to compete successfully in the greenhouse industry.

- **Don't crowd plants.** Tightly spaced plants have to compete for light, expending energy to grow taller rather than producing healthy leaves and robust growth. Again, spindly growth results.

- **Reflect light.** All surfaces in the greenhouse—except those of thermal mass materials—should be painted white, which reflects the maximum amount of light. This is particularly key on the north wall of the greenhouse, which intercepts the majority of winter light. A more intensive strategy is to use light reflectors on the outside of the greenhouse. We comment on this in Chapter 6, as they are typically also combined with insulating shutters.

While these strategies can maximize light in the winter, some growers understandably have higher expectations. You may need to meet a certain crop quality or simply want more variety from the winter garden. If the goal is a cluster of red tomatoes in January, many North American growers will need artificial lighting. Methods for adding supplemental lighting, and doing so energy efficiently, are covered in Appendix 3.

Types of Light and Glazing Placement

Another way to evaluate light conditions is by "percentage of possible sunshine," or simply, cloudiness. This plays a role in where and how much glazing to use in the greenhouse and where to place it. Reason being, clouds affect the *directionality* of light, as shown in Fig. 5.3.

- **Clear days** create intense *direct light*. In these conditions, about 90% of the light comes from a specific area of the sky dome (the south, if in the Northern Hemisphere). This very directional light creates sharply defined shadows and justifies using glazing predominantly on the south side of the greenhouse. (The north side plays a very minor role in light collection.)
- **Overcast days** have uniform cloud cover, which creates *diffuse light*. Cloud cover acts like a lampshade over a bright bulb. Instead of single point in the sky, light emanates from a larger area of the sky dome. Cloud cover ranges from light to heavy, affecting the intensity of light on the ground. Cloudy climates, like those in the northwestern or northeastern US, justify the use of larger glazing areas that are exposed to more of the sky dome.
- **Partly cloudy days** have a combination of conditions. Light alternates from very intense and directional, to less intense and diffuse.

The area of glazing is normally expressed relative to the area of insulated surfaces, or the "glazing-to-insulation" ratio; the goal is to have a balance between the area of glazing and the insulated wall area. Traditional greenhouses are 100% glazed—they don't include any insulation in the walls. Typically, energy-efficient/solar greenhouses have 30%–80% glazed area relative to insulated wall area. In other words, if the total surface area (including all walls

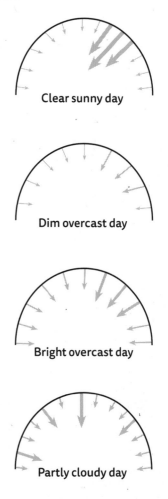

FIGURE 5.3. Light Conditions Based on Cloud Cover. Credit: Adapted from James McCullagh, *The Solar Greenhouse Book*, Rodale Press, 1978

and the roof) amounts to 1,000 sq. ft., the area of glazing would be 300–800 sq. ft. This is a broad range, as it greatly depends on light conditions and when you want to grow. Different climates lend themselves toward different strategies:

- **Mild and cloudy (e.g., Seattle):** Greenhouses in cloudy climates require a larger area of glazing exposed to the sky dome for light levels to be sufficient inside. Optimizing glazing for a specific sun angle or time of year (as discussed below) is less relevant because light is more uniformly distributed throughout the sky dome. Thus, a shallow roof slope and glazing on the south, east and west opens the "aperture" of the greenhouse up to the whole sky. Due to milder temperatures, less insulation is required. Greenhouses may only have the north wall insulated. Or, this is one of the few instances that might justify a fully glazed structure.

- **Cold and sunny (e.g., Colorado):** Given greater light levels and the need to retain heat, greenhouses in cold and sunny environments should be on the lower end of the glazing-to-insulation range. Given the direct light, the majority of the glazing should face south, with an angle optimized for the time of year you want to grow (discussed below). Adding some roof insulation also makes sense. Growers in hot and sunny climates can follow similar advice, with the possibility of adding lower light-transmittance glazing materials and shade cloth.

- **Cold and cloudy (e.g., Connecticut):** This is where it gets challenging. Both glazing for light transmittance and insulation are needed. Generally, we recommend using a moderate ratio and ensuring that the glazing has a good R-value, a material property we'll discuss more below and again in Chapter 6.

When placing glazing, also consider obstructions that might shade the greenhouse. For example, if the east side of the greenhouse is shaded for most of the day, it's a poor area for glazing. That wall is better left insulated.

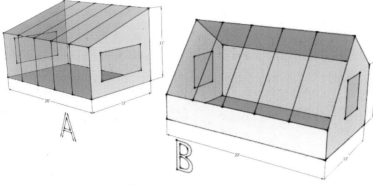

FIGURE 5.4. Glazing-to-Insulation Ratio. Example A has an 80% glazing-to-insulation ratio and is better suited for cloudier climates (hopefully mild ones as well). Example B has a 30% glazing-to-insulation ratio, better suited for climates with direct sun and colder temperatures.

Glazing Materials

Options abound when it comes to greenhouse glazing materials, and sometimes the choices can get overwhelming. To simplify things, we've grouped them into three major categories: rigid plastics, glass and films. Figure 5.5 outlines the most common choices, and their pros and cons. More than the materials themselves, it's important to know how to evaluate a glazing material as the materials are continually changing with advances in manufacturing technology.

Considerations When Choosing a Glazing Material

- **Light transmittance:** This should be one of the top factors you evaluate. Light transmittance is how much of the available light a material transmits, described as a percentage of indoor light compared to light hitting the outside of the greenhouse.
- **Insulation:** Glazing materials are rated by their R-value, which measures their resistance to heat transfer, or simply insulating quality. They can also be rated by their U-value, the inverse of an R-value. Compared to standard insulation or wall materials, glazing materials are very poor insulators: most greenhouse glazings are below R-3. Importantly, the R-value is inversely correlated with light transmission—the higher the R-value, the lower the light transmission. Insulation is added by using more layers of the material with air gaps

FIGURE 5.5. Comparison of Glazing Materials.

		Lifespan in Years	Light transmittance	R-Value	U-Value	Cost[1]	Pros	Cons	Recommended Uses
Glass[2]	Single pane	25+	88–93%	0.9	1.1	Low.	Cheap	Low R-value. Heavy.	Few
	Double pane	25+	75–80%	1.4	.7	Moderate. $5 per sq. ft.	Good balance of light transmission and insulation. Can be sealed well.	Heavy; need framing support	Vertical applications and for view windows
	Double pane, Low-e	25+	60–70%	2–4	.25–.5	High	Enables using glass in roof applications	Cost-prohibitive in large areas	Roof applications if glass is necessary
Polycarbonate	Single layer	10–15	90%	0.9	1.1	Low	Low cost	Low efficiency	Few
	Double wall (6–10 mm)	10–15	80–85%	1.5–2	.5–.6	Low–Moderate. $3 per sq. ft.	Light-weight and durable	Low-R values	Wide range of roof or wall applications
	Triple wall (8–16 mm)	15–20	70–80%	1.8–2.3	.42–.53	Moderate. $5 per sq. ft.	Good balance of light transmission and insulation	Prone to thermal expansion	Almost any roof or wall application, esp. colder climates
	5-Layer (32 mm)	15–40	50%	4	.25	High. $7 per sq. ft.	Very insulating	Low light transmission	Climates with harsh and sunny winters; areas used for sitting or working
Acrylic, Films	Acrylic, double wall (16 mm)	20–30	80–90%	2	.5	Moderate. $5 per sq. ft.	Same as 16 mm polycarb; longer lifespan	Same as 16 mm polycarb	Almost any roof or wall application
	Polyethelene film	2–4	90%	0.83	1.2	Low. $0.01 per sq. ft.	Cheap	Short lifespan. Poor thermal performance	Hoop houses, row covers, seasons extension, or double layer for year-round growing

[1] For the rigid plastic materials, we give the cost of the material plus the attachment system, and other necessary components. In other words, the total material cost, before installation

[2] Glass: assuming placement in a vertical application, not susceptible to breakage

in between, but each layer reflects and absorbs some light. Thus, choosing a glazing material is often a balancing act between light and insulating quality.

- **Cost:** Consider the upfront cost, but also the lifetime costs, based on how often the material will need to be replaced if damaged.
- **Durability and UV stability:** Susceptibility to snow, wind and hail is an important factor, particularly for harsher climates. Additionally, plastic glazing materials will degrade with exposure to UV rays, some much faster than others.
- **Transparent vs. diffuse:** These two terms are often misused. When glazing is *transparent*, it allows a unidirectional ray of light through. You can clearly see detailed objects through transparent materials, like glass. However, most glazing materials are *translucent*: they refract light into many rays (also called scattering, or diffusing the light). A plastic milk carton is translucent—it allows some light through but you can't clearly see through it. Studies have shown that crops grow better under diffuse light conditions because when light is refracted into many rays light, it penetrates deeper into the leaf canopy. Thus, growers prefer diffuse materials, though transparent ones offer nicer views of outside, making the greenhouse feel open.
- **Thermal expansion:** Another factor is how the material reacts to dramatic temperature fluctuations (which occur daily in a greenhouse). Materials will expand and contract with temperature swings and shift over time. This, in turn, causes air infiltration and water leaks. The measurement of this quality is called the "coefficient of expansion," and it is usually specified on a materials spec sheet. If a material has a high coefficient of expansion, this needs to be considered when installing the material; special screws or attachments that allow for shifting should be used.
- **Sealability:** How well can the material act like a sleek marine mammal? No, we don't mean that kind of seal. "Sealability" is how we describe the ability of a material to seal tightly to the frame of a greenhouse. The connection between the glazing and the frame is a major area for potential air infiltration, and, thus heat loss in a

greenhouse. Sealability depends on the coefficient of expansion and how the material is attached to the greenhouse frame. A poor seal results in poor energy performance.

Rigid Plastics

Rigid plastics include polycarbonate, fiberglass and acrylic. These are major players in the hobby greenhouse market today. They're popular for good reasons: they're moderately priced, durable, lightweight and easy to install.

Polycarbonate

Within rigid plastics, polycarbonate is probably the most common in current greenhouses. It has a relatively long lifespan (15+ years), and can be rated for any wind or snow load.

There is a huge range of polycarbonate options, if you find the right distributor (see Further Reading at the end of this chapter). Single-layer polycarbonate is transparent and thin enough that it can be bent over a curved frame. (Thus, this category includes *semi-rigid* plastics.) It is still durable and resistant to hail, but it has an R-value of less than 1. For that reason, we don't recommend it unless you have a hoop house application.

The most common products are 2-layer or 3-layer polycarbonates, also called double or triple wall. These provide decent levels of insula-

Double layer (8 mm) R-value 1.6

Triple layer (16 mm) R-value 2.4

5 layer (32 mm) R-value 4

FIGURE 5.6.
Polygal
Polycarbonate.

tion and high light transmission. You can achieve much more insulation by using more layers, such as a 5-layer material that is rated R-4. However, higher insulation comes at a cost in terms of light transmission: the 5-layer material only transmits 50%–60% of light.

Polycarbonate is also available with different tints or colors that diffuse and/or reduce light. A milky-colored version, called opal, reduces light significantly. This would be a good choice for plants that like dimmer, diffuse light, such as orchids. It could be useful for greenhouse roofs in very hot climates like the Southwest.

All these products are lightweight and easy to install, provided the right support and components. They come in panels, typically 4' or 6' wide, and a variety of lengths. Either a plastic or an aluminum track is used to attach it to the frame. We use an aluminum track called Mega-Lock, a durable product that gives the roof an attractive finished look. In addition, there are a number of other parts required: special screws to allow for shifting from thermal expansion; a vapor barrier for the edges; end caps; and gaskets. (These can be sourced with the material from a greenhouse or plastics distributor.) Given these accoutrements and the high-quality material itself, most polycarbonate roofing systems are in the mid-range of cost, at $3–$7 per sq. ft.

FIGURE 5.7.
Polycarbonate Installed. "Bubble washer" screws, gaskets and an end cap are some of the components needed to install polycarbonate. Credit: Ceres Greenhouse Solutions

Historically, the disadvantage of polycarbonate was that it was susceptible to UV damage which caused it to turn yellow and become brittle over time. However, its UV resistance has greatly improved in recent years. Most products now come with a 10–20-year warranty.

Fiberglass and Acrylic

Acrylic and fiberglass were the main rigid plastic glazings used before polycarbonate dropped in cost and improved in performance. Commonly known by its trade name "Plexiglas," acrylic is similar to polycarbonate in many ways. It can come in a multi-walled form, making it a good choice for roof and wall applications. It can also be bent over a shallowly curved frame. Acrylic is slightly less impact resistant than polycarbonate—it will shatter more easily—but is still very strong as a material. (Acrylic has 17 times the impact resistance of glass. Polycarbonate has 250 times the impact resistance of glass.) We recommend making a decision based on manufacturers and costs in your area. It is likely that polycarbonate will be more widely available and cheaper to procure.

Fiberglass is made by embedding shards of glass fibers into a plastic resin. Most varieties are opaque; it's used in storage tanks, sports helmets, boat hulls, etc. Translucent varieties (typically a milky color) can be used for greenhouse glazing; however, they have lower light-transmission. Be aware that it is also flammable, and it often has a rough texture that can trap dirt and further reduce your light transmission. Though fiberglass used to be common in greenhouses, today it has mostly been replaced with polycarbonate.

Glass

Glass has one clear advantage, literally speaking: it is transparent and provides a view window. For residential greenhouses, this makes the greenhouse feel open and spacious, much like a sunroom. If you plan on using your greenhouse as a sitting/relaxing space, we highly recommend incorporating some glass to avoid feeling like you are sitting in a small box, though a nice box it may be.

There are two types of glass products: single-pane glass (often called plate glass) and Insulated Glazing Units (or IGUs). Both have an extremely long lifetime if used in the right applications, as they are impervious to UV damage. Single-pane glass is a single sheet of glass that can come in many sizes (in standard-sized sheets as replacement glass for storm doors, custom cut from a manufacturer, or small panes from older greenhouses). Plate glass has very low insulating qualities and is much heavier than plastics (requiring a sturdier frame), so we don't recommend it. If you want something thin and transparent, one of the single-layer clear, rigid plastics would be a better choice.

A better alternative to plate glass is Insulated Glazing Units, or IGUs, the windows used in homes. Double-pane types consist of two layers of glass with an air gap in between. Layers greatly increase the R-value, usually R-2.5 for a clear double-pane window—making it an energy-efficient option. They are encased in a variety of materials—like vinyl or fiberglass—and come in a range of sizes. You may be surprised at the sizes of IGUs you can order from the hardware store at a reasonable cost per square foot. There is quite a variety of operable and fixed windows in a range of frames types. Vinyl is the lowest cost, but it also experiences the greatest thermal expansion.

The primary disadvantage with glass is that it can shatter. For that reason, it is best used in vertical applications, not in angled walls or roofs where it is exposed to falling objects. If used in a roof, you will probably need to use tempered or laminated glass, which is typically required by building codes. These added elements prevent the glass from breaking into dangerous shards; however, they also increase costs tremendously. For these reasons, we typically avoid glass in roofing applications. Glass is generally more expensive and much heavier than its plastic counterparts, and it requires more framing if used in the roof. While many IGUs have warranties, they may not apply to use in a greenhouse application. The glass itself has an extremely long lifetime, but the seal of the window can break, causing it to fog up.

Given its pros and cons, we think glass has a place in residential and educational greenhouses; however, it should be used sparingly and

Q&A

Should I use low-emissivity (low-e) or argon-gas filled windows?

Window options go far beyond clear glass. There are thousands of variations of residential windows that increase energy efficiency. These add films or tints to the glass, or gases between the panes. Low-e windows, for instance, reflect much of the IR and UV wavelengths, while allowing the visual spectrum through. (In turn, they reflect IR wavelengths back into the room, increasing efficiency.) While that in itself is not harmful to plants (since they don't use the IR or UV wavelengths for growth), some coatings can cut out part of the PAR spectrum, diminishing growth. As an anecdote, once we went to visit a high-tech conservatory that used a highly energy-efficient tinted glass. Although it looked relatively clear, it transmitted less than 40% of PAR light due to the coatings.

Light transmittance varies by individual product. If considering a more advanced window, talk to a supplier that specializes in greenhouse glass. Also check the visual transmittance (VT) of the window, which can serve as an indicator for PAR light transmittance (since they span the same range). In general, we've found low-e and argon-gas-filled windows are acceptable if you want to upgrade to more energy-efficient glazing. We don't recommend tinted windows—they cut out PAR light.

---

cautiously. We recommend clear double-pane windows for vertical walls. This provides a high light-transmittance and moderately well insulating material, while keeping the upfront cost down.

In most of our residential projects at Ceres Greenhouse Solutions, we combine a tri-wall polycarbonate for the roof with vertical glass windows on the south, east and west sides. We typically use operable windows on the east and west for added airflow to supplement ventilation. We most commonly use vinyl-framed, double-pane glass windows without advanced energy-efficient tints or coatings.

Plastic Films

Polyethylene

You've probably seen polyethylene film on hoop houses. This thin, malleable plastic is very cheap upfront but has a very short lifespan. It has applications for greenhouses in mild climates and for season extension in colder areas. For year-round greenhouses in most places with freezing winters, we don't recommend it.

First, poly can easily collapse or get torn off in wind, snow or hail. We've seen many polyethylene structures go down in the first year of use, some after only a few months. The film is highly susceptible to degradation from UV rays and becomes brittle, needing to be replaced every few years. If you are considering poly, we recommend using a UV-resistant product, which increases its lifetime some. Even so, having to replace a whole greenhouse every 2–4 years is costly, both environmentally and financially.

A second disadvantage is that the thin film is a horrible insulator. It has an R-value less than 1 and does not seal tightly to a greenhouse frame. For that reason, poly films are often used in double layers with air blown into the cavity in between them. If you do opt for polyethylene, we recommend going this route, even though it requires some additional equipment, like an inflation blower. The air pocket provides some insulation; however, the poly can still tear easily, producing holes that cause the greenhouse to become a sieve.

The pro of poly films is simply the low cost—the reason it is the most popular material in the greenhouse industry today. Polyethylene provides a level of crop protection and season extension without much investment. It can also be sufficient if you only want to grow cold-hardy crops over the winter. (Eliot Coleman is famous for growing mache and other greens in unheated polyethylene greenhouses through the winter in Maine.)

Poly is undoubtedly the most flexible material; it can be used on any structure, no matter the size or shape. A great application is as a season extender covering raised beds or a hoop house in the fall and spring. It can then be taken down during the winter and summer to reduce weathering. Due to the flimsy nature of films, we don't recommend them for year-round applications in a solar greenhouse design.

A new material is ETFE (lovely full name: ethylene tetrafluoroethylene). It is also a malleable plastic film, but much more durable and insulating than poly film. It is becoming more common as an architectural glazing to cover curved buildings with a translucent material. Notably, the biome greenhouses of The Eden Project in Cornwall, England, are glazed with ETFE. Currently, most distributors are in Europe, so sourcing is difficult. Because it is so new, we have not seen enough data or applications to comment definitely on its performance; however, it is one to keep watching.

Calculating the Angle of Glazing

It is one of the most common questions: What is the best angle for the glazing in a solar greenhouse? The optimum angle of glazing depends on the seasons you want to grow in. More specifically, it depends on the seasons in which you want to maximize light and heat in the greenhouse (typically winter for most year-round growers). There are many articles with simple answers to this question floating around on the internet, but seldom do they explain the reasoning behind their rules—and that can lead to some misguided decisions.

A common rule of thumb for calculating the "best" angle of glazing—either in the roof or south-facing glazing—is to take your latitude and add 20 degrees. While valid in its logic, this rule is overly simplified. Moreover, it can be problematic because it creates extremely steep roof pitches. For example, in Denver, latitude 40 degrees, this would create a roof pitch of 60 degrees. Such a steep roof necessitates either a very tall building or altering the geometry of the greenhouse, which can increase costs. We'll evaluate these geometries in Chapter 8; for now, it's helpful to understand how light interacts with glazing based on the angle.

The first thing to understand is that when light hits a material "straight on" or perpendicular to the glazing material, the maximum amount of light is transmitted. The angle of a ray of light *relative* to perpendicular is called the *angle of incidence*, shown in Fig. 5.8. When that angle is 0 (perpendicular) the maximum amount of light is transmitted. (This is about 87% for a single pane of clear glass. The other 13% is reflected or absorbed.) When the angle of incidence increases, more light is reflected instead of being transmitted.

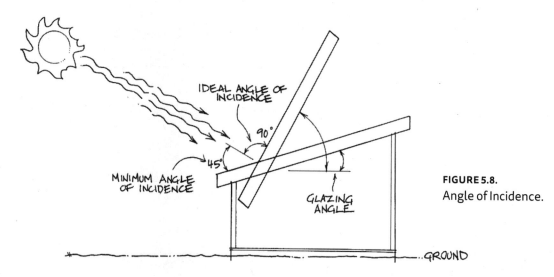

FIGURE 5.8.
Angle of Incidence.

Importantly, this relationship—between the angle of incidence and light transmission—is not linear. For polycarbonate and some other materials, there is almost no decrease in PAR light transmission when the angle of incidence is less than 45 degrees. In other words, as long as the glazing is within 45 degrees of perpendicular, the decrease in light transmission is insignificant. This is true for all glazing materials: going from a perpendicular angle to an angle of incidence of 45 degrees typically reduces light by only 1%–5%. With triple wall polycarbonate, for example, there is virtually no change (about 1%) in light transmission when the material is perpendicular to the sun and when it is tilted 45 degrees away from perpendicular. Thus, as long as you stay away from very shallow roof angles, there is little effect on light transmission due to reflectance. For roof angles at 40 degrees latitude, then, the minimum threshold is greater than 15 degrees.

To find the range for your location, simply modify the common rule of thumb to account for the great deal of tolerance in glazing angles:

1. First follow the common rule of thumb to find the perfect angle of glazing in the walls and roof: take your latitude and add 20 degrees. This will create a perpendicular, or nearly perpendicular, angle to the sun in the winter months, yielding the maximum light transmission.

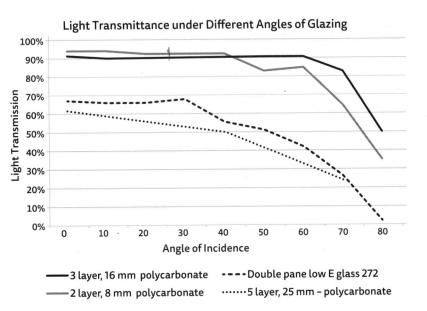

We use 20 degrees because this is a good average for the winter months. Some designers use a higher number, like 25 degrees, which creates a perpendicular angle to the sun at winter *solstice*. We recommend using an average, not choosing the angle when the sun is at its absolute lowest point; this allows for better light transmission over the whole winter season. If you want to optimize for an entirely different time of year, see this basic process laid out in more detail in Appendix 2.

2. Second, subtract 45 degrees from the number in step #1. This is the practical angle of incidence that yields sufficient light levels during the winter. Any shallower than this and light dramatically starts to decline.

That's the first half of the story. The second half has to do with the area of glazing that is exposed to the sun, often called the area of incidence. You can see how the area exposed to the sun changes by comparing the shaded areas in Scenarios A and B in Fig. 5.11. Scenario B (representing a 45 degree angle of incidence) will receive about 30% less light than

Scenario A because it has a smaller area directly exposed to the sun. In this way, the angle of glazing (either roof or walls) can significantly impact how much light the greenhouse collects.

It is important to realize, though, that this also changes the building's dimensions. Think of a greenhouse as a collector of solar energy (both light and heat). A building that is 20' tall (such as in Scenario A) will collect much more light than one that is 10' tall because it is a larger collector. Though a tall greenhouse with a larger glazing area will receive more light, it may go against building codes or neighbors' preferences. Likewise, if the south wall is slanted at an angle (instead of vertical), it creates a larger area of incidence for winter light, but also requires a larger footprint and foundation. In short, while it is possible to increase light with a larger building, this in turn can increase complexity and cost.

So how do you factor in the building dimensions when choosing the angles of glazing areas, particularly the roof? We recommend first choosing dimensions based on your budget and other practical factors. Building an extremely tall building just for the sake of light collection is not practical. For the height, we recommend a maximum wall height of 9'–12' for residential greenhouses. Once you have determined the dimensions, and thus the size of the solar collector, you can then find the

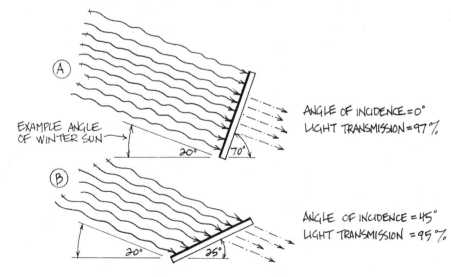

EXAMPLE ANGLE OF WINTER SUN

20°   70°

ANGLE OF INCIDENCE = 0°
LIGHT TRANSMISSION = 97%

20°   25°

ANGLE OF INCIDENCE = 45°
LIGHT TRANSMISSION = 95%

FIGURE 5.10.
Glazing Angles.

range of glazing angles that allow for maximum light transmission using the simple steps given above.

Finally, note that this discussion revolves around glazing angles but light transmission also depends on the glazing material you choose as well as shading factors in the greenhouse, like roof beams. Material choices/shading factors tend to have an even greater effect on light transmission, decreasing light by 30% or more. Most commercial growers strive for total light transmission levels of 70% or more.

Other than light, several other factors should be considered when it comes to choosing glazing angles:

- **Greenhouse shape:** Different greenhouse shapes—what we call geometries—allow for different glazing areas and angles. Some greenhouses allow for one steeply pitched glazing surface. A single steeply angled roof uses less glazing than both a vertical wall and roof. Using less glazing creates a more efficient structure because it reduces the most inefficient material; however it tends to increase the complexity of the build. Other greenhouses use two glazing areas: a vertical wall combined with a glazed roof. We'll compare these options more in Chapter 8, Greenhouse Geometries.
- **Climate:** In cloudy climates, precisely calculating glazing angles is less relevant. Solar radiation is diffused across the sky and comes from many angles.
- **Summer heating:** While steeper glazing angles maximize light in the winter, they also deflect more light in the summer—generally advantageous for preventing overheating of the greenhouse. Given the high angle of the sun in the summer, much more light will be reflected off the glazing. In other words, the summer sun creates a very severe angle of incidence, which reduces transmission.

Pitch	Angle
4:12	18.4
6:12	26.5
8:12	33.7
10:12	39.8
12:12	45.0

Pitch vs. Angles

The building industry describes roof slopes as fraction of rise over run, using increments of 12 as the run. Once you determine the angle of glazing, you can easily equate this to a pitch. Some examples appear in the table here.

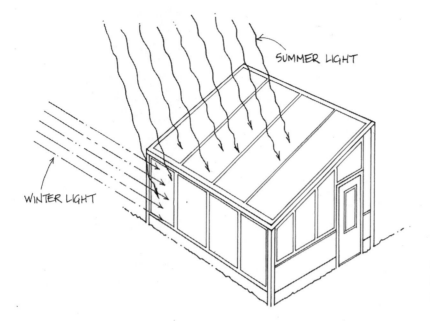

SUMMER LIGHT

WINTER LIGHT

FIGURE 5.11.
Summer vs. Winter
Light Exposure.

- **Snow loads** Finally, steeper pitches of the roof allow it to shed snow more easily. This allows for the use of thinner framing members— less structure is needed to support snow loads—which reduces costs and enhances light.

Using Multiple Glazing Materials

As mentioned above, solar greenhouses can have one or multiple glazing areas. Using two different glazing materials allows you to be particularly strategic about controlling light and heat transmission.

As the angle of the sun changes through the year, from low in the winter to high in the summer, different glazing areas play a lesser or greater role in light transmission. In the winter, the vertical or near-vertical surfaces will be the predominant collectors of light and heat because they have a greater surface area exposed directly to light. In the summer, glazing areas with shallower slopes are the predominant collectors, as shown in Fig. 5.11.

To maximize light in the winter, use a high-light-transmittance glazing on south-facing vertical walls, (the areas that will absorb the most light during the winter). Even though these have lower insulation values, they can be very valuable for heat gain. In residential greenhouses, double-pane glass windows serve nicely here. They have a high light transmission (>95%), and are nice as view windows.

A lower light transmittance material on the roof will help control overheating during the summer. This also has the advantage that the material will also have a higher insulation value in the area of greatest heat loss. Consider thicker, more insulating (and more light diffusing) polycarbonates, which have the advantage of being resistant to hail.

Regardless of your greenhouse design, understanding how the angles and materials affect light transmission affords you some control over the growing environment.

Shade Cloth

In addition to full ventilation, adding shade cloth to the glazing during warmer months is extremely helpful to keep the greenhouse from overheating in the summer (if you intend to use your greenhouse in the summer). To estimate whether or not shade cloth is needed, consider the average summer temperatures in your climate. If it is 80°F (27°C) outside, at best the greenhouse will be 80°F, and likely much warmer (unless you use mechanical equipment like swamp coolers). That climate will likely overheat plants. Shade cloth helps reduce heat gain, to make daytime temperatures more manageable.

A natural (and attractive) shading method is to use plants trellised on the greenhouse walls and roof. Morning glories, passion fruit, hops, beans...any fast-growing vine that can tolerate heat and sun can be grown on the south side of the greenhouse or across the roof. While a greenhouse covered in flowering morning glories is an enticing image, it obviously requires some time and patience for the plants to grow.

Shade cloth is a simple, low-cost, and highly effective alternative. The most common products are made out of woven polyethylene. They come in a variety of shading factors, rated by how much light they block, such as a 25% light-blocking material. Which you choose depends on the intensity of the summer sun and what you are trying to grow.

Shade cloth is most effective when placed on the *outside* of the greenhouse. Installed on the interior, it still lets heat build up between the cloth and the glazing. When applied outside, it must be secured using tie-downs to avoid having it being torn off in wind. Mechanical systems can also be used as backup cooling, if necessary, in hot climates. Evaporative coolers and misting systems are other aids to summer growing in hot and dry climates.

Takeaways

- Typically solar greenhouses use 30%–80% glazing area relative to insulated wall area.
- When considering glazing materials, evaluate their durability, insulating quality and light transmittance.
- Glass is best used in vertical applications; rigid plastics like polycarbonate are good for most applications due to their light weight, durability and balance of light transmission/insulating quality.
- Steeper pitches for glazing on the south wall and the roof are generally better from a performance standpoint: they maximize light transmission in the winter, allow less light in the summer, and shed snow more easily. The downfall with steeper pitches is that they create impractical heights and are costlier to build.
- A range of roof angles will provide close to maximum light transmission. Stay away from very shallow roofs as this dramatically reduces light transmission in the winter. Consider snow loads, ease of building and other factors when choosing a roof slope, not just light.
- If using multiple glazing materials, use a high light-transmittance glazing on south-facing vertical walls and a lower light-transmittance material on the roof, which will help control overheating during the summer.

Further Reading
Measuring Light and Light Levels for Crops

Torres, Ariana P. and Roberto G. Lopez. "Measuring Daily Light Integral in a Greenhouse," Department of Horticulture and Landscape Architecture, Purdue University, May, 2012, extension.purdue.edu

Runkle, Erik. "Light It Up," Michigan State University Extension, hrt.msu.edu
/energy

Biernbaum, John. "Hoophouse Environment Management: Light, Temperature,
Ventilation," MSU Horticulture, hrt.msu.edu

Glazing Materials

Most plastic products are sold through greenhouse retailers or regional distributors. You can search for distributors in your area by going to the manufacturers'
websites, or searching online for plastic distributors in your area.

For more information on energy-efficient window options see The National
Fenestration Rating Council: nfrc.org

Case Study: Ecotope
The Parabolic Aquaculture Greenhouse

384 sq. ft. Experimental Greenhouse
Ecotope Group
Arlington, Washington

How do you grow year-round when light levels are only one tenth of the levels recommended by commercial growers? That is the situation for many growers in the Pacific Northwest climate. From the late 1970s through early 80s, a group of researchers addressed this issue with an inventive greenhouse located on a Washington farm. Davis Straub, one of the designers, summed it up: "The idea was to aim the sunlight into the fish pond at the back of the greenhouse to be a storage heat source that would steady the inside greenhouse climate." The north wall was framed as a parabolic curve so that it would catch the winter sun and direct it downward toward plants and a large in-ground fish tank housed on the north wall. The 4,800-gallon fish tank served as the thermal mass for the structure, passively regulating temperatures.

The greenhouse also included a "passive cooling stack" consisting of a glazed column with a black interior. The stack was meant to heat up during the day. As the hot air inside rose, it exhausted out the top, drawing cooler air into the greenhouse; this created a strong convection current that greatly accelerated passive ventilation. At its peak, the cooling stack exhausted air at a rate of 1,600 cubic feet per minute (the rate of an exhaust fan). It worked extremely well to keep temperatures below 85°F (29°C) through the summers. In the winter, the large fish tank was a suitable source of thermal mass. On average, indoor temperatures stayed 14 degrees above outdoor nighttime temperatures over the first two years.

While some elements were successful, the north wall reflector was not very effective at concentrating and directing solar energy. Reflectors work best when there is a unidirectional ray of light to reflect. Given the low-light levels and cloudy climate (diffuse light conditions), a greenhouse with a larger glazing area would have been a more logical choice. Davis Straub agreed,

Manufacturers

Polycarbonate: Brand names are Polygal, Lexan and Palram (Sunlite and ThermaGlas).

Acrylic: Acrylite.

Endnotes

1. "Measuring Daily Light Integral in a Greenhouse," Ariana P. Torres and Roberto G. Lopez, Department of Horticulture and Landscape Architecture, Purdue University; extension.purdue.edu

commenting: "The parabolic greenhouse was a clever design, but not a very successful one given the cold and cloudy climate and the cost. I'm quite certain that there are cheaper ways to build such a greenhouse and better locations for it." However, innovation tends to happen in increments. As one of the early trials of a solar greenhouse, the Ecotope greenhouse advanced both greenhouse design and aquaculture/aquaponic growing strategies.

FIGURE 5.12. Ecotope Greenhouse in the 1980s. Credit: Ecotope, Inc.

CHAPTER 6

Controlling Heat Loss: Insulation

Minimizing heat loss is one of the key differences between a smart, solar greenhouse and the standard energy-guzzling one. Though both have similar amounts of solar gain during the day, a solar greenhouse is designed to trap this heat overnight by reducing heat loss through the walls, roof and ground as much as possible. Insulation allows the greenhouse to maintain warmer temperatures at night and through cold winter days.

How Heat Loss Occurs

The purpose of insulation is to reduce conductive heat losses through the surfaces of the greenhouse. Conduction—the movement of heat through a material—is the primary reason the greenhouse cools down at night and on cold days. As shown in Fig. 6.2, the vast majority of conductive losses occur through the glazing materials, which are much less insulating than the walls.

The second largest source of heat loss in the greenhouse is typically air infiltration: when warm air leaks out the greenhouse through cracks or gaps, or cold air seeps in. Here, too, the glazing area is the biggest culprit because most glazing materials create air gaps as they expand and contract under temperature swings. To compare, most homes have one half to one full air exchange per hour. (Very efficient homes have one third air exchange per hour; that is, they only lose a third of their air volume every hour). Traditional greenhouses are notoriously "leaky" structures and have upwards of two air exchanges per hour.

It's impossible to create a perfectly sealed environment, but it is possible to greatly reduce air infiltration with some simple, cheap methods. It's critical to caulk all cracks and seams in the greenhouse after it's built. Given proper caulking, and "tight construction," we find it possible to keep air exchanges to less than one per hour. Secondly, avoid windy locations when siting the greenhouse. Wind increases air infiltration because it blows through cracks. It also creates pressure differentials around the greenhouse, spurring more air movement.

Convection—the movement of heat as air warms up and rises— plays a corollary role in heat loss. As air expands and rises, it passes across the glazing, where conductive heat losses occur.

More importantly, convection is responsible for a salient characteristic of greenhouses: extreme temperature differentials between plant levels and the peak. As hot air expands and rises, it collects at the highest point in the greenhouse, causing the top zone to commonly be 40 degrees F warmer than plant level. This phenomena allows for passively cooling the greenhouse during the day. Hot air at the peak can be vented out, creating a convection current that draws cool fresh air in.

We point out these terms because, while much of the conversation revolves around insulation—reducing convective heat loss—it is imperative to know this is only part of the picture. You can install super-insulated walls, but if the entire volume of air in the greenhouse is exchanged 2–3 times per hour due to air infiltration, the insulation becomes much less relevant. Insulation must be combined with a well-sealed environment for the greatest effectiveness.

Understanding R-Values:
How Much Insulation Is Enough?

The universal metric for rating insulation materials is the *R-value*. It measures a material's resistance to thermal conduction, or its insulating quality. The higher the R-value, the more insulating. R-values dominate any discussion of insulation because they are a simple value that can be applied to any material. However, what is less discussed is that R-values are only part of the equation that measures the total heat loss through a wall or surface, which can be defined as:

$$\text{Heat Loss} = \left(\frac{1}{\text{R-value}}\right) \times (\text{surface area}) \times (\Delta T)$$

R-value = resistance to heat transfer of a material
ΔT = the difference between the inside and outside temperatures in degrees F.

Importantly, the relationship between R-values and heat loss is not linear, as shown in Fig. 6.1. That fact has major implications when choosing an insulation strategy. We say "strategy" because you naturally have choices about where and how much to insulate. Fig. 6.1 shows that heat loss plummets as you move from an R-1 to an R-2 but changes relatively little when increasing from an R-20 to an R-21. Though in both cases the R-value increases by 1, the first upgrade will have a huge impact on total heat loss, the second comparatively little.

The implication is that money is much better spent adding some insulation to low-R-value surfaces like glazing, rather investing greatly in "super insulated" walls with very high R-values. Due to this non-linear relationship, we recommend ensuring that all glazing materials have a decent R-value (at least R-2).

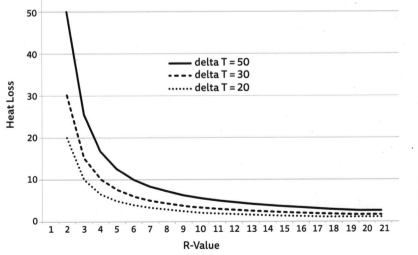

FIGURE 6.1.
Relationship between R-Values and Heat Loss.

FIGURE 6.2: Effect of R-values on Heat Loss. Variables: 12' × 20' greenhouse, located in Boulder, Colorado, with a 50% glazing area-to-wall ratio (415 sq. ft. of glazing in the roof and walls, and 413 sq. ft. of insulated wall [for scenarios 3–6]). Slab insulation assumed at R-10. Air infiltration assumed 1 per hour. Calculations made using the Heat Loss Calculator on builditsolar.com.

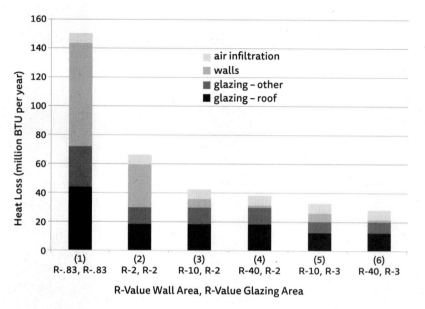

To further illustrate this point, Fig. 6.2 compares different combinations of R-values for the glazing area (in both the roof and on the walls) and for the insulated wall area. We use a hypothetical greenhouse, and keep other variables the same, changing just the R-value of each surface.

The first scenario represents a traditional uninsulated greenhouse with polyethylene or single-pane glass glazing (R-0.83) on all sides and the roof. There is no insulated wall area. The second assumes the same structure has a double layer of glazing (R-2) on all surfaces. The third adds R-10 insulation on the north wall, and some on the east and west. Now the greenhouse has what we would call a 50% glazing-to-insulation ratio, about half the area is insulated with an R-10 wall and the other half with R-2 glazing. From there, we alter the two variables slightly, changing the insulated wall area to R-40 (run #4), or changing just the glazing area to R-3 (run #5). The last represents the most efficient structure, combining an R-40 wall with R-3 glazing.

The change in total heat loss demonstrates the payback of upgrading glazing materials to a decent R-value. Adding the second layer of glazing (going from R-1 to R-2) cuts heat loss by over 50%. Adding the insulated wall reduces total heat loss by another 35%. From there, the payback becomes less straightforward. Going from an R-10 to an R-40 wall only creates heat savings of 10%. Whether this investment is worth it depends on your goals for the greenhouse, the current cost of heating, the cost of the insulation material, and your climate. To evaluate the effect for your area, we recommend performing your own quick analysis using an online heat loss calculator, such as "The Home Heat Loss Calculator" on builditsolar.com. Hand calculations are also possible, but are becoming less relevant given the ease and functionality of online calculators. (See Further Reading for more resources.)

We flesh this out because a common mistake we see is the devotion of lots of money and effort to creating super-insulated walls (R-40 or greater) while at the same time using very poor glazing materials. The justification is usually, "I need a thin glazing for light transmission." However, adding a second layer of glazing is usually only a 10% reduction in light. Given the enormous energy savings from that extra layer, in our opinion, the increased temperatures and performance are worth it.

What is the difference between R-values and U-values?

The U-value is the reciprocal of the R-value (U=1/R). It indicates the ability of a material to conduct heat; therefore, a lower number is better. It's most commonly used for glazing materials, like window units, and in heat loss calculations.

Insulating the Glazing

As demonstrated by an exercise like Fig. 6.2, the lion's share of heat loss in a greenhouse occurs through the glazing. The good news is it is easy to improve the insulating quality of glazing by using multiple layers of material. Layers create air gaps, which insulate and transmit light.

We recommend a minimum of two layers of glazing. This can be done by layering multiple sheets of material (such as two layers of polyethylene), or using a multi-layer material (such as a tri-wall polycarbonate or double-pane windows). The added cost of a thicker multi-layer material is not very big, and the energy savings are huge, making this a no-brainer when it comes to solar greenhouse design.

A second strategy is simply to minimize the area of glazing as much as possible. (This has less to do with actually insulating, and more to do with adjusting the geometry of the greenhouse, the topic of Chapter 8). The geometry of some greenhouses minimizes the glazing area while creating an angle that allows for year-round light collection. In other words, they minimize the glazing-to-insulation ratio and maximize the annual light gain on the glazing surface. This is an excellent tactic for areas that have harsh winters and sufficient light. In cloudier climates, you will have to rely on a larger glazing area.

Moveable Insulation

Highly insulated glazing materials are out there, such as a 5-layer polycarbonate or triple-pane, low-e windows. Typically, though, they are either too low in light transmission, or too expensive for large areas. R-values under 5 are the norm for glazing—and hard to get around. To accommodate for that fact, a common strategy among solar greenhouses builders is the use of moveable insulation to cover the glazing at night. This creates the best of both worlds: the glazing is a light collector during the day, and becomes like an insulated wall at night.

The challenge with these systems is simply the moveable part. Automation requires tracks, motors and a system customized for your greenhouse. There are not many easy kit products out there, so automated moveable insulation systems are either quite expensive or very time intensive. They are justifiable for large commercial greenhouses, but rarely so for residential greenhouses. Thus, moveable insulation usually requires an active operator to move it in the morning/night.

The strategies vary from very simple to complex. At the most basic level, you can install an extra layer of a glazing material or bubble wrap—commonly sold for greenhouses—during the winter months. This is left up, but comes with the challenge that water can easily get between the two materials and fog up the glazing, reducing light transmission.

The next step up is moveable curtains or shutters. Curtains are a popular method for large commercial greenhouses. They're made out of aluminized fabrics, which reflect thermal radiation back into the greenhouse. Large-scale systems are automated with motors and tracks to pull the curtain across the glazing. This is one case when a cost-effective automated solution is possible, though it is only justified at a certain scale. A number of studies show that thermal curtains can cut nighttime heating loads by 50%. They can also be used as a shading element during the summer months.

While a great solution for commercial growers, unfortunately there are not many easily installed curtain systems for small greenhouses on the market. Creating one typically involves purchasing a curtain material (found through a greenhouse supplier) and building a custom track system out of wire cables. Do not use fabrics or insulating shades made for residential windows—they will quickly mildew and mold in the high humidity of the greenhouse. The best fabrics are sold from greenhouse retailers and contain an aluminized fabric, which reflect thermal radiation back into the greenhouse at night.

An additional strategy for fitting curtains to your greenhouse is to use track systems intended for something else. Try looking for "shade curtains" or "black-out systems" for parts and supplies when building a curtain system. See the resources in Further Reading as well.

A major challenge with curtains of all types is getting a good seal against the wall or glazing. Without a seal, warm air will be drawn between the curtain and the glazing, negating much of the effect of the curtain. Magnetic strips or Velcro can help create seals, but they are pretty inelegant solutions.

Instead of curtains, rigid insulation in the form of shutters can be manually moved and placed over the glazing at night. These provide more insulation and better seals—if done right. The basic versions are made from rigid foam board insulation, sized to fit the opening of the window or glazing. Called pop-in shutters, these are placed snugly in the opening and secured with a latch, bungee cord, or wood board. The disadvantage is that the boards are awkward to move and store when not in use. We recommend using them as a "once in a while" backup in cases of extreme weather and storing them in the garage for most of the season.

FIGURE 6.3.
Pop-in Shutters.
Credit: Ceres
Greenhouse Solutions

The next level involves creating permanent shutters that are hinged to the outside of the greenhouse and close over the glazing at night, as shown in Fig. 6.4. The most common design is to hinge shutters below the window and flip them up to cover the glazing, though there are many variations. Permanent shutters have a unique benefit: they can also be used as light reflectors during the day. The topside of the shutter can be covered with a reflective metal or plastic. It can be positioned at the correct angle, relative to the angle of the sun, to direct more light into the greenhouse during the day. This gives you the capability to both increase the total amount of light during the day and turn your greenhouse into a tightly sealed, insulated box capable of withstanding extremely harsh conditions.

This shutter/light shelf combination is a brilliant concept in theory, but it's beset by some practical challenges. First, from a materials perspective, it is hard to create a large shutter that is both lightweight and able to achieve torsional stability (making the shutter rigid and exactly plane to close evenly against the window). Shutter frames can be made of welded metal or very thick wood. The first requires some custom

FIGURE 6.4. Moveable Insulating Shutters. A motor and pulley system closes these shutters/light reflectors over the windows at night, creating a highly insulated window at night. Credit: Ceres Greenhouse Solutions

welding (expensive, if you don't do it yourself); the latter creates very heavy shutters (difficult to lift and secure by hand). But you can add a pulley system to assist in opening/closing them. To go a step further, the shutter can be automated using a timer and a motor. Keep in mind these are custom solutions that require time and money, appropriate for some growers who are doing the building themselves, but impractical for many others.

The final scenario—creating an automated shutter and light reflector—is the holy grail of moveable insulation. This was one of our early endeavors at Ceres, but it turned out to be much more difficult than expected. In addition to the challenges mentioned above, snow and wind loads create further complications for building stable, automated shutters. All the challenges were solvable, but added significant cost and engineering time. We constructed shutters out of extruded aluminum frames and rigid polyiso insulation at the center, backed by plywood. Taking into account installation, an attachment system, and automated control, this custom system added a few thousand dollars to the residential greenhouse. The cost would be lowered if designing and installing the system yourself. Again, moveable insulating shutters are excellent options for those who enjoy building and engineering custom solutions—and all the time/tinkering that entails—but difficult for "hands-off" growers.

Insulating Underground

When considering insulation, most people think about the walls, glazing and roof. Less considered, but equally important, is the ground. Heat loss occurs through the floor of the greenhouse just as it does through the walls, because the natural temperature of the topsoil is about the same as the outdoor air temperature (i.e., freezing in areas with freezing winters).

When we say greenhouse floor, we mean the ground plane. Many greenhouses don't use flooring materials; they have bare ground so that plants can be grown in the soil. (Even if growing in raised beds, we recommend connecting the beds to the soil underground.) We don't recommend creating an insulated floor, per se. Rather, underground insulation should extend around the *perimeter* of the greenhouse, as shown in Fig. 6.5, leaving the greenhouse connected to the soil underground.

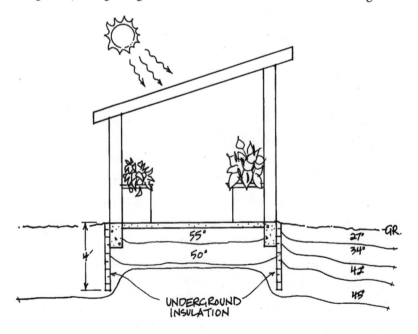

FIGURE 6.5.
Effect of Underground Insulation. Insulating around the perimeter creates a "thermal bubble" of warmer soil temperatures directly beneath the greenhouse.

The purpose of underground insulation is two-fold. First, it insulates the indoor soil from the surrounding topsoil, reducing heat loss and keeping the soil warmer year-round. Secondly, it connects the greenhouse to the stable temperatures of the soil deep underground. Once insulated, the soil underground becomes useful thermal mass. It passively absorbs and stores heat, creating stable temperatures year-round—what we call the "thermal bubble" underneath the greenhouse. This prevents the soil inside from freezing. Moreover, it helps stablize indoor air temperatures by integrating a large source of thermal mass.

The reason for underground insulation stems from the fact that anywhere in the world, the soil deep underground stays warmer and fluctuates less than air temperatures (as shown in Fig. 6.6). Across the US, average soil temperatures below the frost line are around 40°F–50°F (4°C–10°C). The exact range varies by climate. You can find data for many areas by using the National Resources Conservation Service (NRCS) database, called the Soil Climate Analysis Network (wcc.nrcs .usda.gov/scan/).

Installing Underground Insulation

There are two strategies for installing perimeter insulation: vertically (straight down); or horizontally (sloping it away from the edges of the

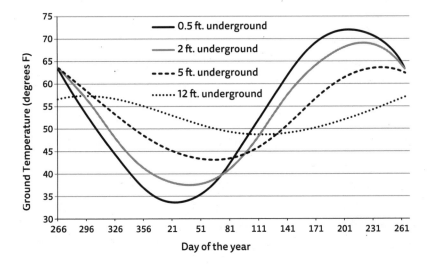

FIGURE 6.6. Effect of Depth on Soil Temperature. (Example: Temperatures vary by climate)

Q&A

How deep should insulation be installed?

Like many decisions, this comes down to striking a balance between cost and performance. The greater the distance of insulation, the more stable soil temperatures will be underground. We usually insulate to a depth of 4'. In most climates, soil temperatures 4' underground start to even out and stay warmer year-round, as shown in Fig. 6.6. Four feet is also the standard width of insulation boards, so it keeps material costs down. In climates that have harsher conditions (say, colder climates like northern Canada or Alaska, or very hot climates like the southwestern US), it is worthwhile going to greater depths (e.g., 6'). In climates that have mild conditions (for example, parts of California), insulation may not be needed at all, or it can be used to very shallow depths (e.g., 2'). It all depends on your soil temperatures. You can look yours up using the NRCS database for US locations (www.wcc.nrcs.usda.gov/scan/).

greenhouse). Which is easier depends on your site preparation, and whether or not you need to excavate. Excavation usually depends on your foundation type and if you are installing an underground heat exchanger like a ground-to-air transfer (GAHT) system, the subject of Chapter 13. Foundation types are discussed more in Chapter 9. Thus, the best method for your greenhouse—vertical or horizontal insulation—will become clearer as you progress with your greenhouse design.

Vertical insulation is more logical if you are already planning to excavate underneath the greenhouse (either for foundation reasons, or if installing an underground heat exchanger). In the vertical method, boards of rigid foam insulation surround the foundation of the greenhouse, as shown on the last page of the color section. We recommend overlapping two layers of 4' × 8' insulation boards around the foundation. The excavated soil is then back-filled and allowed to settle before the greenhouse is built.

If you don't plan on excavating underground, horizontal insulation is probably easier. Also known as wing insulation, the boards of insulation slope away from the edge of the greenhouse, 2'–4' away on all sides, as shown in Fig. 6.7. Unlike vertical insulation, this is best installed after

Recommended Insulation: Underground

We normally use two layers of rigid foam board insulation (extruded polystyrene) to achieve an R-12 to R-14 layer of insulation around the perimeter of the greenhouse.

the greenhouse is fully constructed on-site. To install, a backhoe or excavator excavates an area sloping away from the greenhouse, starting about 9" deep and extending to about 2' deep at the edge. Ideally, a layer of gravel is first laid on this soil (to help with drainage underneath the insulation). The boards of insulation are placed on this sloping surface. The slope is critical to allow for drainage away from the greenhouse. On top of the insulation goes another layer of gravel (well-draining soil can also be used).

In either method, extruded polystyrene foam board (commonly called blue board or pink board) is the standard insulation for below-grade applications; it can withstand higher levels of moisture. We normally install two layers of boards, providing a total of R-10 to R-14 around the perimeter of the greenhouse. Boards should overlap so there aren't gaps between boards.

A recent concern has emerged: currently, all types of polystyrene are treated with a fire retardant called HBDC, a known toxin. Research is going on now, including investigations into whether or how HBDC can leach into soils and have effects there. As a result of these concerns, at Ceres we are looking for safer and more sustainable alternatives for underground insulation. Possibilities include mineral wool or perlite, but these are still limited in availability. Extruded polystyrene is still the norm.

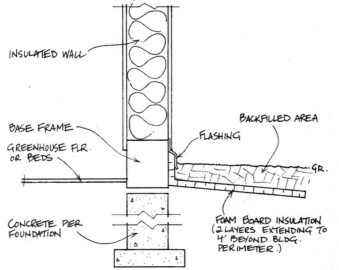

FIGURE 6.7.
Horizontal/Wing Insulation. Rigid foam board is laid at a slight slope to allow for drainage, extending out from the base of the greenhouse.

Regardless of the method, underground insulation is relatively cheap to install and can have a major impact on all greenhouses. Thus, it's one of our first suggestions for greenhouses of all types, whether small residential greenhouses with no foundation, or commercial-scale greenhouses that a full foundation.

Insulating the Walls and Roof

Insulating the walls of a framed solar greenhouse is much like insulating a house: the same materials options are available to you, though some are better than others for a greenhouse.

Batts and Rolls (Fiberglass and Mineral Wool)

Fiberglass is ubiquitous in every type of construction for a very simple reason—it is cheap and easy to install. However, you get what you pay for. The "field performance" of fiberglass is much lower than its rated R-values because it can easily wick moisture. For that reason, we don't recommend it for greenhouse applications (because moisture is a constant factor).

A similar, but more effective material is mineral wool batts. Mineral wool includes rock wool and slag wool (different materials formed into the same mineral wool batts). In contrast to fiberglass, these batts are semi-rigid: they come in boards that are firmer and thus easier to cut and get a secure fit. The R-values are also significantly higher, at 3–4 per inch. Finally, mineral wool is more water resistant than fiberglass, making it a viable candidate for insulating greenhouse walls. The drawback is that at the time of this writing it is not as readily available as fiberglass (it must be sourced through distributors or certain retailers), though this is changing quickly as its popularity—particularly among green builders—is increasing.

Rigid Foam Board (Polystyrene and Polyiso)

Rigid foam board is another very common type of insulation that can be used in walls and underground. Foam board has a higher moisture resistance than fiberglass, giving it better performance over time. There are three types: extruded polystyrene (XPS), expanded polystyrene (EPS), and polyiso. Because that is a lot of "polys" we normally use their non-technical names as references.

Extruded polystyrene is normally called "pink" or "blue" board due to brands' characteristic colors. It is very common both for walls and below-grade applications. The advantage of pink/blue board is that it has high insulation values for its cost, and it is very moisture resistant. The latter makes it a very good material for using in humid environments like greenhouses and also underground. It's commonly used to insulate around the foundation of a greenhouse. Unfortunately, it is not the "greenest" material—making it creates ozone-depleting compounds and, as mentioned above, it currently includes a fire-retardant chemical called HBDC in the US.

Using a different manufacturing process, companies also make *expanded* polystyrene (EPS). Commonly called "beadboard," these rigid boards are made of a conglomerate of white beads (the same material as Styrofoam cups). Though more sustainable than its extruded counterpart, EPS has a lower R-value per inch and worse moisture resistance. For those reasons, it is less common than pink/blue board for greenhouse applications.

Polyiso (full name polyisocyanurate) is another type of rigid plastic insulation. It has a higher insulation value, usually R-6 per inch. Though the total cost per board is higher for polyiso boards, the cost per R-value is the lowest among rigid insulation products. For that reason, this tends to be our top choice when insulating the walls of the greenhouse, as it has the greatest insulation value per dollar spent. It absorbs moisture slightly more readily than polystyrene, but still has adequate-to-good moisture resistance. Polyiso boards come with a reflective foil backing. The predominant brand name is R-Max.

Recommended Insulation: Walls

We recommend polyiso rigid foam board to insulate the framed (nonglazed) walls of the greenhouse. Polyiso provides high R-values for its cost, while also being water resistant. In addition to insulating in the wall cavity (between the studs) we highly recommend using a layer of polyiso behind the studs. The board of polyiso is installed so that it covers each stud, underneath the exterior siding. This reduces heat loss through the wood framing (also called thermal bridging). The conductive losses of uninsulated studs significantly reduce the R-value of the wall overall. We also recommend insulating around the edges of the board with insulating foam sealant (brand name: Great Stuff) to seal any gaps.

Rigid foam boards (of any material) come in 2' or 4' widths, typically 8' long. You can get them 1, 2 or 4 inches thick. So the overall R-value can be anywhere from R-2 to R-16, depending on the thickness and rating. Recommended R-values vary by climate, as we'll discuss below.

Blown-in Materials

Blown-in insulation can consist of cellulose, fiberglass or mineral wool. All are made out of some recycled content (either newspaper, glass shards, or industrial waste products), processed into slurry that is blown into the wall cavity. These are less commonly used in greenhouses. Though they provide high R-values, the installation is usually cost-prohibitive. Blown-in methods require significant preparation and setup by professional installers. The small insulated wall area of most greenhouses makes this impractical cost-wise. Blown-in insulation may have a place in very large commercial greenhouses; it depends on cost estimates provided by local installers.

Insulating the Roof

So far, we've discussed using better glazing materials on the roof. To go one step further, you can insulate the top section of the roof that is not needed for light collection. The top portion of a glazed roof does little to directly illuminate plants. Rather, light coming through the top section hits the top of the back wall, due to the angle of sun (shown in Fig. 12.4). Although some light is reflected off the wall, the contribution to the plants below is minimal (unless plants are reaching all the way up the wall). While not a necessity, roof insulation adds to the overall energy efficiency of the structure, by both reducing unnecessary heat gain during the day and heat loss at night. If left uninsulated, the glazing here experiences the highest rates of heat loss.

In a stick-framed greenhouse, the insulation is best installed between the rafters under the roof, using the same materials and methods as the walls. More information on finding the right length of roof insulation is given in Chapter 12.

Minimum Levels of Insulation

Like many quantitative recommendations in this book, the "perfect" insulation level is unique to you; it depends on your outdoor climate and your goals for what you want to grow. Other factors include the heating/cooling systems in the greenhouse; the greenhouse design; whether or not you will be relying on backup heating; the cost of installation; and the cost of backup heating. To take all those variables into account and produce the optimal metric requires some advanced energy modeling in a program like Energy Plus. Energy modeling is usually above and beyond the needs of the average grower (though consultants/greenhouse designers can help). However, calculating the heat loss of the structure is a practical step. A heat loss calculation can't tell you the "best" level of insulation, nor can it give you the predicted indoor temperatures. It can compare the effect of different insulation strategies on total heat loss and thus heating costs, creating an analysis like the one shown in Fig. 6.2. This can tell you the relative effect of different upgrades, such as going from an R-2 to an R-3 glazing material or an R-20 to R-21 wall. Several online heat loss calculators are available; a few are listed in Further Reading at the end of this chapter.

Some people are inclined toward this kind of quantitative analysis; others prefer a simple, albeit more generalized, recommendation. Figure 6.8 gives basic recommended R-values of walls and glazing areas according to growing zone. (You'll need to reference your growing zone; see Chapter 3). We say "basic" because these are broad recommendations; they aim to provide good insulation levels to keep the greenhouse above freezing for most of the year. They assume that the greenhouse has a form of efficient heating, thermal storage, or climate control. They are infused with a bit of common sense, since we recognize the level of insulation also depends on cost, and there are diminishing returns to adding more and more insulation.

FIGURE 6.8. Recommended Insulation Levels.

Growing Zone	Walls	Glazing
1–5	R-20 to R-40	R-3 or greater
6–9	R-14 to R-20	R 2–3
10–13*	R-10 to R-14	R 2–3

* These climates do not experience freezing conditions and year-round growing outdoors is common. In very hot climates, insulation is useful for keeping the greenhouse from overheating rather than overcooling.

Takeaways

- Heat loss occurs through conduction through the walls and roof of the greenhouse, predominantly the glazing areas. To minimize heat loss, the north and some of the east and west walls should be well insulated; the glazing should include at least two layers of material with an air gap in between. The exact level of insulation depends on your climate and growing goals.
- Air infiltration is a big contributor to heat loss; it should be minimized by caulking all cracks and seams in the building after it is complete.
- In most climates, the greenhouse can greatly benefit from insulation installed around the perimeter of the greenhouse underground. This keeps the soil beneath the greenhouse warmer, and allows it to tap into the stable temperatures of the earth deep underground.

Further Reading

Heat Loss Calculations

"Home Heat Loss calculator" provided at builditsolar.com/References /Calculators/HeatLoss/HeatLoss.htm

Many resources give step-by-step instructions on how to perform a heat loss calculation by hand instead of using an online calculator or website. We recommend the steps laid out by Rob Avis in his free ebook, *Passive Solar Greenhouses: A Do-It-Yourself Guide*, available at vergepermaculture.ca

Insulation Levels and Strategies

National Resource Conservation Services database, the Soil Climate Analysis Network, (www.wcc.nrcs.usda.gov/scan/). You can find soil temperature data here.

McCullagh, James. *The Solar Greenhouse Book*. Rodale Press, 1978. Recommended insulation levels are given.

Langdon, William K. *Moveable Insulation: A Guide to Reducing Heating and Cooling Losses in Your Home*. Rodale Press, 1980. A good guide to designing/ building insulating curtains and shutters for glazing.

ENERGY Star, Recommended Levels of Insulation, energystar.gov. Recommended R-values for standard construction.

US Department of Energy, Energy Saver, energy.gov

CHAPTER 7

Ventilation

When we started designing greenhouses, we thought the primary challenge was keeping them warm enough. Greenhouses operate primarily through the colder months right? We didn't anticipate that keeping the greenhouse *cool* enough was an equally great challenge. Fortunately, though, it's one that can be easily remedied with proper ventilation and thermal storage methods. This chapter explains how to equip a greenhouse with sufficient ventilation via passive vents or electrically operated exhaust fans.

Why Ventilate?

Here in sunny Colorado, the temperature of a closed greenhouse can easily soar over 100°F (38°C) on a clear day—and that's not even during the summer months. The greenhouse effect is surprisingly powerful. An anecdote can best explain: One of our clients had purchased a kit plastic greenhouse and left it closed and unused except for storing some foam boards of insulation inside. After a couple of months, she pulled the boards out to find them completely disfigured. The greenhouse got so hot, they melted like a stick of butter in the microwave.

So, you can expect a greenhouse to get very warm during the day. Most plants get stressed at temperatures over 90°F (32°C). Even with thermal storage methods to absorb excess heat during the day, most greenhouses need ventilation to provide cooling for much of the year.

First, we should clarify some terms. "Cooling" describes anything aimed at lowering temperatures in the greenhouse, including shade cloths, evaporative coolers, and misting systems. The easiest, most energy-efficient method (by far) of cooling a greenhouse is exchanging the greenhouse air with cooler outdoor air ("ventilation"). In addition to cooling, ventilation serves other vital functions that keep the greenhouse healthy, such as circulating air, adding CO_2, and reducing humidity.

Ventilation Strategies

Ventilation methods can be categorized into those that require electricity (active systems) and those that don't (passive systems), summarized in Fig. 7.1. All provide some level of automation, via a thermostat or a vent opener. An additional possibility is to manually open windows and doors in the greenhouse. This can be a nice backup method during the summer, but we don't recommend relying on it. The greenhouse can heat up and cool down quickly, and it's hard to be present at just the right the time to open and close windows. Often, the greenhouse will just need a short period of ventilation, requiring multiple trips to adjust windows. Plus, the manual method makes it hard to take a vacation.

FIGURE 7.1. Ventilation Methods.

Method	Active or Passive	Pros	Cons
Passive solar vent openers	Passive	No electric usage; automated.	Difficult to precisely control; issues with operation under snow and wind; more difficult to build and install.
Exhaust fans	Active	Very reliable and precisely controlled. Easy install if electricity is available.	Require electricity; add noise to the greenhouse.
Solar vent (attic) fans	Active (but self-powered)	Same as above; don't require wiring or a power source for the fan.	Only operate when sun is shining; low CFM rates.
Heat Recovery Ventilators (HRV)	Active	Provide energy-efficient ventilation without overcooling the greenhouse.	Very expensive; low CFM rates.

FIGURE 7.2.
Bayliss Solar Vent
Opener. Credit: Superior
Autovents (Sturdi-Built
Greenhouses)

Passive Vents

Passive vents rely on devices called *solar vent openers* to open and close vents. These contain a wax cylinder that expands when heated, opening the vent. When the temperature drops, the wax contracts and closes the vent. The cylinder can be set to operate at a certain temperature (within a set range).

The second element that facilitates passive venting is natural convection. As air heats up, it rises and is exhausted out of vents located at the top of the greenhouse. Intake vents located as low as possible on the greenhouse wall draw cooler, fresh air into the greenhouse. The greater the distance between the lower and upper vents, the greater the airflow. There are a variety of possible configurations, shown in Fig. 7.3.

FIGURE 7.3.
Vent Placement.

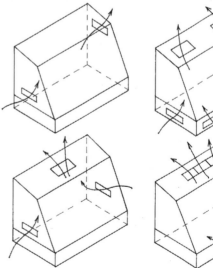
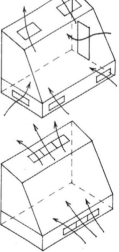

FIGURE 7.4.
Consider snow buildup when placing intake vents. Credit: Dirt Craft Natural Building

There are a number of things to keep in mind when locating vents:

- **Wind directions at your site:** Ideally, the exhaust vent should be on the leeward (downwind) side of the greenhouse. Wind blowing against an exhaust vent will impede airflow.
- **Snow and ice buildup:** Many people consider building exhaust vents in the roof. It's possible, but much more complicated given that snow and ice can build up on top of a vent and prevent it from opening, possibly breaking the vent opener. Additionally, roof vents create exposed edges where water runoff can leak down into the greenhouse. Roof vents require more careful construction, normally raised above the roofline so that water runs around the lip of the vent, not down into cracks (the same way skylights are installed).
- **Snow shedding:** For intake vents, snow buildup at the front of the greenhouse can be a problem as shown in Fig. 7.4. A vent trying to open into a mound of snow will break. You can disable some vents

when they are not needed in the winter; build overhangs into the roof; or go above and beyond, creating special awnings above intake vents to protect them from snow

- **Interior layout:** When locating intake vents, sketch out the location of growing beds or any equipment inside that could block a vent. Typically, intake vents are located in the south wall above the growing beds. In winter, though, plants can be "shocked" with bursts of freezing cold air when the vent opens. To avoid this, consider disabling some vents during the winter. The east and west corners are also good places for vents.
- **High winds:** Vents protruding from the greenhouse have the tendency to act like sails; they can get torn off in wind gusts. Most vents have a "safety cord," but these are not fail proof. If you are in a very windy location, consider building a windbreak around the greenhouse. Locating intake vents as low as possible and putting exhaust vents on the leeward side of the greenhouse keeps them out of the strongest winds.

As you can see, the effects of wind, snow, ice and water penetration make designing and building vents more complicated than it first appears. Think carefully about how the elements will affect vents to ensure their longevity. We find the easiest locations for exhaust vents are high on the vertical north wall of the greenhouse; intake vents are best low in the east or west corners of the greenhouse.

Vent Construction

A vent is a simple concept in theory—it's an opening with an operable covering—but the devil's in the details. It's hard to get something to both move easily and close tightly. The edges of vents are prone to air infiltration (heat loss) and water penetration. The best way to seal vents is with gaskets—either foam or urethane—where the vent frame connects to the frame of the greenhouse. Both intake and exhaust vents should be sealed as well as possible, as these can be a major area of air infiltration (heat loss).

Having quality, well-sealed vents normally requires building your own. There are a few turn-key systems available from greenhouse retailers like FarmTek, but the options are limited. Additionally, kit vents are relatively cheap systems that will not seal well to the greenhouse. They are only available in a limited number of small sizes and may be hard to accommodate in your greenhouse build.

Thus, most growers build their own custom vents if using a passive venting strategy. Vents can be framed out of metal or wood. Wood is more practical unless you have welding experience. Large, wood-framed vents can get very heavy, so vents are typically moderate sizes, less than 10 sq. ft. in area and sized to fit between the studs or framing members. Sometimes, they span two stud bays with a header on top.

The specifics of vent construction depend on where it is located. If intake vents are part of a glazed wall, the same glazing material, like polycarbonate, can be used as the center of the vent. Longtime passive solar greenhouse designer Cord Parmenter (see the case study at end of

FIGURE 7.5.
Roof Vent.
Credit: Raven
Crest Botanicals

this chapter) frames intake vents with aluminum extrusions sold from greenhouse retailers. He cuts the extrusion to length, inserts the polycarbonate in the center, and pop rivets these together (using screws is also possible).

Most builders place intake vents in an insulated knee-wall or on the side walls. If a vent (either intake or exhaust) is built into an insulated wall, it can be framed out of wood and insulated with rigid foam board insulation like polystyrene or polyiso. The vent is then sided with the same greenhouse siding material. The vent is hinged to the greenhouse frame, and the solar vent opener is attached per the device's instructions.

Vents are always best when built into vertical or shallow-sloped walls. This reduces issues with snow, ice and water. In these cases, the construction is conceptually straightforward: build frame, insert glazing or insulation, hinge to frame, and install opener. Exhaust vents located on the roof require some added steps to prevent water leaking down through the cracks around the vent. They are raised slightly above the roof plane so that water runs around the vent, not over the cracks. Flashing covers the lip. *Add-On Solar Greenhouses and Sunspaces* by Andrew Shapiro has good diagrams showing multiple ways of building quality roof vents.

Even when built yourself, vents are prone to air infiltration and heat loss. To accommodate, we recommend sealing and insulating any vents that are not needed in the winter. Often, just one pair of vents (intake and exhaust) can suffice to provide the small amount of ventilation needed in winter. The rest can be insulated for the season by disconnecting the vent opener and inserting a block of foam board insulation, like polyiso or polystyrene, in the vent cavity.

Choosing a Solar Vent Opener

There are a variety of solar vent openers available from online retailers. The main brand names are Gigavent, Univent and Bayliss. When evaluating options, consider the lifting capacity of the opener. It's crucial to get an opener that can lift your vent, which will be heavy if wood framed. (The exact weight will vary by where it's installed as well. Roof vents pushing *up* will need more force than vertical vents swinging *out*.) Gigavent makes a robust opener that can lift up to 60 lbs.

Secondly, check the "maximum opening size" of the opener—how wide it will open the vent. The effective area of a vent is not equal to the size of the vent itself, but only its length times the distance that it opens. An opener that protrudes farther allows more airflow; however, be cautious of wind and the sail effect.

Finally, check product warranties and general durability. Solar vent openers are often damaged in high wind and snow events. A more expensive but robust opener is a probably better investment in the end. The wax cylinder, or "tube," of an opener will need to be replaced after a few years—these inevitably degrade with use—but most companies offer replacement tubes.

Sizing the Vent Area

The typical rule of thumb when sizing vents is that the total area should equal 15%–20% of the square footage of the greenhouse.[1] Half of that area should be intake vents, and the other half exhaust vents. Doors and operable windows can be included in the vent area—they will contribute if they are open. We don't recommend relying on manually operated doors and windows for too much of your vent area, since they are not automated. They are useful to add extra ventilation during the summer. As described below, for warmer months doors and windows can be left open full-time.

Planning the size of your vents usually requires some give-and-take calculations. To illustrate, say you have a 12' × 20' (240 sq. ft.) greenhouse. The above rule of thumb would dictate 24–48 sq. ft. of total vent area. If your greenhouse is built with framing 16" on center, it would make sense to have each vent about 16" or 32" long.[2] Say you chose 32" and plan to build each vent 24" high. Thus, each vent is a little over 5 sq. ft. (We will round to 5 sq. ft. for simplicity.)

To achieve the recommended area you could have:

- Three intake vents and three exhaust vents (a total of 30 sq. ft. of vents)
- Four intake vents and four exhaust vents (a total of 40 sq. ft. of vents)

From there, you can evaluate your climate to estimate the final area. If you live in a climate with intense sun and warm temperatures, the larger vent area would be wise. A cool and cloudy climate would call for the smaller area. Or, if you only plan on using the greenhouse in the winter, you can make the vent even smaller—a total of 20 sq. ft. Other considerations are the spacing and layout. If the greenhouse were a larger footprint, it may call for spacing out smaller vents in more areas to achieve uniform airflow.

As you can tell, several considerations go into this process, but generally it's easy predict and use your best judgment to determine size and vent spacing. If unsure on vent area, we recommend going with a larger size. It's always possible to close and seal vents if they are not needed, but it's quite difficult to add more venting later if the greenhouse is overheating.

Pros and Cons of Passive Venting

Passive vent openers have a couple other quirks you should be aware of before relying on them. First, while you can adjust the cylinder to open at different set temperatures, they are not as precisely controlled as a thermostat. This is because the mechanism of control is located at the vent, not at the center of the greenhouse. Rather than registering the temperature where the plants are, the vent responds to the temperature directly around it. Exhaust vents can stay open too long in the evening because they experience warmer temperatures at the peak of the greenhouse. Thus, precious heat can be lost just when greenhouse needs it most. Intake vents, on the other hand, are inundated with cold air from the outside when they are open. They are prone to closing too early, leaving the greenhouse vulnerable to overheating. These are not insurmountable issues; though, solar vents simply require some time and fiddling to get them to open at the right time.

Though not as precise as fans, passive solar vents are the easiest automated and passive venting method. If you do not plan on adding electricity to the greenhouse, they are an essential tactic. Growers that are looking to use the space as a relaxing/sitting area or a classroom, may favor the quiet, passive venting solution.

Motorized Window Openers

Instead of solar vent openers, you can open vents (or windows) with a motorized opener. This is similar to passive solar vent openers in that the actual ventilation is passive—facilitated by convection currents—but the vent opener requires a small amount of electricity as the vent opens. (Exhaust fans, in contrast, may need power for a few hours each day.) Thus, it's quite easy to power a motorized vent opener off a small PV panel connected to a thermostat.

The downside is that motorized vent openers are usually more expensive, in the range of $200 each—a significant cost if you are using them for multiple vents. Most temperature-controlled motorized openers are designed for skylights and awnings; they are not greenhouse-specific products. Finding a product that is right for your windows setup requires a little more research and customization.

The plus is that they can be more precisely controlled. If using multiple on different windows they can be set to open at slightly different temperatures so that the greenhouse cools down gradually, reducing the shock of cold air on plants in the winter. Motorized openers can also create a tighter seal, and be used with a more insulating window instead of a custom-built vent.

Motorized options are a good fit if you have operable windows and want to take advantage of these as your vent area. We recommend integrating them with awning windows (which swing out) versus slider windows (which slide sideways). With a slider window arrangement, the motorized vent opener often has an arm that sits on the window and restricts how far it can open. Combined, an awning window and an automated opener is a great plus for efficiency and aesthetics but much more costly than vents or fans.

Exhaust Fans

Exhaust fans offer precise control and reliability. A thermostat located at plant level activates the fans, which run until the greenhouse is cooled. The drawbacks are the electric usage and noise. For residential greenhouses, we find the latter is typically more of the determining factor.

As we'll show below, the energy costs of running fans are minimal. The larger cost is simply bringing electricity out to the site. If fans are the only electric appliances, it is more cost-effective to use a passive system or the solar-powered exhaust fans described next. For other growers, especially commercial ones, exhaust fans are essential to achieve easily controlled, reliable ventilation.

Using exhaust fans requires three components: a fan, an intake shutter and a thermostat, all shown in Fig. 7.6. The fan exhausts hot air from inside the greenhouse to the outside. An intake shutter is a simple vent covered with louvers; it's needed to provide a source of incoming air when the fan is running. These can be motorized, or they can be allowed to freely flap open when the fan turns on. The motorized versions allow for more controlled airflow (the flaps don't blow open in a strong wind) but are more expensive.

FIGURE 7.6.
Exhaust Fan Kit.
Credit: Greenhouse
Megastore.com

Like passive vents, intake vents and fans can be the Achilles' heel of an energy-efficient greenhouse during the winter. Both parts are aluminum, a highly conductive metal that can easily sap heat to the outside. To reduce heat losses, we recommend covering and insulating fans and intake vents during the winter when ventilation isn't needed daily. We find this is best done by building a simple moveable insulating shutter, as shown in Fig. 7.7. A simple wood frame with insulation at the center can be hinged to the outside of the fan and vent. (Make sure to unplug the fan when you close the shutter so it does not run while the shutter is closed). The frame can be easily propped open again when temperatures are warm enough to allow ventilation. Thus, it's an easy and low-cost way to curb heat losses from leaky, aluminum fans and intake vents in the winter.

FIGURE 7.7.
Exterior Insulating
Shutter Over an
Exhaust Fan.
Credit: Ceres Greenhouse
Solutions.

Exhaust fans should be placed high on a wall so they exhaust the hottest air out of the greenhouse. Intake shutters should be located lower down on the opposite wall to take advantage of natural convection, drawing air up and across the greenhouse. Ideally, the intake vent should be on the *windward* side of the greenhouse and the exhaust fan on the *leeward* side of the greenhouse (if you have consistent wind patterns). For example, in Boulder, Colorado winds predominantly come from the west, so we typically put exhaust fans in the northeast corner of the greenhouse and intake vents low on the west side.

Fan Sourcing and Sizing

Fans are rated by how much air they can move in units of cubic feet per minute (CFM). The standard rule of thumb for sizing fans is to create at least 1 exchange of the entire air volume every 1–2 minutes. That means if your greenhouse has a volume of 2,400 cubic feet (say it is a 12' × 20' with average height of 10'), a good fan size would be 1,200–2,400 CFM.

Some growers size the fans smaller than this recommended airflow rate if they have other, more efficient means of storing heat during the day, like thermal mass. While that's a logical and appropriate step, we typically stick to the standard recommendations when choosing fan sizing. If the fan is sized too small, you risk seriously overheating the greenhouse, and possibly needing to replace the fan at some point. Plus, there are few disadvantages with a larger size. Currently, a 12" fan with maximum flow rate of 1,100 CFM costs about $145. A 16" with maximum rate of 2,950 CFM costs $175. The power usage is not hugely different between the two. If unsure which to pick, we recommend spending the extra $30 to size the fan larger in case there is a greater need for cooling than you expect. The fan will simply run less often if there is less of a cooling load.

Many fans have variable speeds, which allow you cover range of flow rates. Of course, consider your climate (sunny or cloudy) when choosing a fan within the recommended range. For a specific calculation, an energy analysis is needed. This can be done in professional software program like Energy Plus, or via consultation from a greenhouse designer.

Considerations when buying a fan are the energy usage (rated in amps), noise level (decibels), and size. You can estimate electrical costs of running fans by multiplying the energy usage with how often you expect it to run. For example, say you select the 2,400 CFM fan for your residential greenhouse. It uses 2.5 amps. You first need to convert this to Watts (a measure of power) using the "WAV" equation, Watts = Amps × Volts. (Note most fans run on 115 or 120 Volts.)

In our experience, fans typically a couple to a few hours per day when the sun is out. (This can be more accurately estimated with an energy analysis.) You can then estimate the annual energy usage and the cost of running an exhaust fan by following some simple calculations:

2.5 amps × 120 volts = 300 watts

Assuming the fan will run an average of 3 hours per day over the year:

300 watts × 3 hours = 900 watt-hours, or .9 kWh used per day
.9 kWh per day × 365 days = 328.5 kWh per year
328.5 kWh per year × $0.15 per kWh (estimate) = $49 per year

From that sample calculation, you can see that using a fan in a residential greenhouse is not very expensive, but that all depends on how often it runs (and your interpretation of expensive). Other measures—thermal storage and strategic shading of the greenhouse—keep dependence on the fan down.

The fan you purchase will stipulate the required rough opening to fit in the wall. Fans are either hard-wired or plugged to a dial thermostat that in turn is wired or plugged into an outlet. We prefer the plug-in options, as the outlet can be used for something else if the fan is not in use. The thermostat should be set a few degrees below the maximum temperature you want the greenhouse (i.e., 85°F [29°C] if your threshold temperature is 90°F [32°C]). Plug the fan into thermostat, and you have reliable ventilation and cooling.

FIGURE 7.8.
Solar-Powered Vent Fan.
Credit: solaratticfan.com

Solar-powered Exhaust Fans

Solar-powered exhaust fans are a recent addition to the array of ventilation options, for greenhouses at least. These small exhaust fans include a solar photovoltaic module that powers the fan. The PV panel is direct wired to the fan, so it can only run when the sun is shining. They are popular options for cooling attics, and thus you can find most products by searching the more popular name, "solar attic fan."

Most solar-powered fans are a reasonable cost ($200–$500) but typically have low flow rates (most are under 1,000 CFM). They are intended for attic applications, not sunny greenhouses. A possible variation is to purchase a larger DC (direct current) exhaust fan and create your own custom system, adding a small direct-wired solar panel to run the fan. We'll return to this tactic in Chapter 16, Powering the Greenhouse.

A major caveat to these fans is that they only work when the sun is shining. For many climates, this doesn't impede their use because the vast majority of ventilation is needed on sunny days. This can be problematic, though, for very warm climates (if the greenhouse needs cooling on overcast days). Another drawback is that most aren't manufactured for greenhouse (i.e. humid) environments. Because they are relatively new to the market, we can't comment on their lifespan in a greenhouse, and we recommend checking the warranty of the product.

Solar powered fans are an excellent option if you want to otherwise forgo electricity. They are also a good means of providing supplemental venting in the winter. The demand for ventilation in the winter is only a fraction of that for the rest of the year, so a small solar fan can take the place of larger passive vents (which can then be sealed up and insulated in the winter).

If purchasing a solar-powered fan, make sure that it includes, or can be wired with, a thermostat. Without a thermostat, the fan will run whenever the sun is out, including on cold winter days, so you could end up with an overly cool greenhouse. Just like a standard exhaust fan, you want a solar-powered fan to kick on only when the greenhouse needs ventilation.

Season-specific Strategies

When it's 30°F (−1°C) outside and the greenhouse is struggling to stay above 40°F (4°C), the last thing you want to do is blast it with freezing cold air. Cooling is occasionally needed in the winter, depending on the intensity of sunshine, but often it is not. At times when cooling is not needed, the greenhouse still needs some humidity control, air movement, and enough CO_2 to fuel growth. At these times, the benefits of ventilation are at odds: you don't want to overcool the greenhouse, but do want fresh, dry air. There are several backup strategies we recommend to help maintain a healthy greenhouse environment when ventilation isn't advisable. These include:

- **Air circulation** is essential for a healthy growing environment, particularly in the winter. We'll return to this vital part of greenhouse growing below.
- **Heat Recovery Ventilators (HRVs)** are air-to-air heat exchangers, devices that can provide dehumidification and CO_2 supply without overcooling the greenhouse. They preheat incoming air using the outgoing air of the greenhouse. The two air channels pass by each other, without mixing, in a box installed in the wall. HRVs work much like an exhaust fan except that, instead of drawing in 30°F (−1°C) air in the winter, the HRV prewarms the air to say, 45°F (7°C). In this process, it also lowers the relative humidity of the

greenhouse. As the incoming air is warmed, it's able to hold more moisture, reducing humidity levels.

The major drawback is the cost. Most units are over $600 before installation (many much more). Additionally, they tend to have lower flow rates than exhaust fans, so an additional exhaust fan is typically required. HRVs are good extra step for growers who want to provide winter ventilation and humidity control as energy-efficiently as possible. They also have applications in attached greenhouses in which humidity control is more of a concern.

- **A Ground-to-Air Heat Transfer (GAHT) system** is the topic of Chapter 13. A GAHT system stores heat from the greenhouse in the soil underground. As it does so, it circulates and dehumidifies the air. This is possible because the warm, humid air from the greenhouse is circulated through the cooler soil during the day. The air is cooled underground, and, when it reaches the dew point, the water vapor condenses underground. Perforated pipes allow the water to drip into the soil, near plants roots. In this way, a GAHT system helps take humidity out of the air and moves condensed water into the soil. Though its primary purpose is temperature control, these are great corollary benefits.

- **Dehumidifiers** forgo ventilation and just provide dehumidification within a closed environment. They are rare in residential applications, but reasonable for commercial greenhouses that require a tightly controlled growing environment. More expensive machines simultaneously dehumidify and heat the greenhouse through a process of phase changes.

Another strategy is to ventilate manually on warmer days in the winter to replenish CO_2. Say there is a sunny mild January day; it would be a good time to open up the windows fully for a bit and flush the greenhouse with fresh air.

During the spring and the fall, automated venting systems are particularly helpful because venting requirements change day by day and hour by hour. In the summer, we recommend keeping the greenhouse fully open. If nighttime temperatures are warm enough to allow it, all windows, doors and vents can be open full-time. Allowing the green-house to cool at night helps any thermal mass reset and be able to absorb more heat the next day.

Air Circulation and Humidity

High humidity is often the biggest surprise for first-time greenhouse growers/builders. They don't plan for it, and often use materials and components not made for humid environments, which in turn quickly degrade or break. Furthermore, high relative humidity can be a major detriment to a healthy growing environment. The reason involves basic processes of plant growth.

A key part of photosynthesis is *transpiration,* in which plants suck up water and nutrients from the soil to new growth at the top, where photosynthesis occurs. The mechanism for this works a lot like a straw: plants use negative pressure to draw up water from the soil. What creates the negative pressure? Evaporation. As water evaporates from leaves, the plant naturally sucks more water and nutrients up from the soil. Evapo-ration is also the plants' cooling mechanism. If you ever wonder why greenhouses are so humid, just look at every green surface—during the day plants are constantly emitting water vapor as they grow.

Without air movement, the air directly around the leaf becomes saturated, and evaporation slows, since it is more difficult for plants to release water vapor into very humid air. As a result, photosynthesis slows, nutrient uptake suffers, and a whole host of things go wrong with the plant. (Blossom drop in tomatoes, for example, can be caused by high humidity.) Furthermore, in stagnant conditions, plants use up the CO_2 in their vicinity, leaving a thin layer of CO_2-depleted air around the leaf. Equally important is the fact that most molds, mildews, bacterial pathogens, and insects thrive in stagnant, humid air.

The ideal relative humidity levels are in the range of 40%–70% for most growers. This is easily achieved for much of the year when the greenhouse is ventilated some. However, the winter it becomes more challenging, hence the winter-specific strategies above. The good news is that these issues are easily remedied by controlling humidity and pro-viding regular, gentle air circulation, as discussed below.

It's easy to test humidity levels with low-cost digital thermometers that include a humidity reading. Keep in mind that humidity is relative to the air temperature, which changes over the course of the day. Warm air holds more water, so relative humidity is always lower during the day. At night, the temperature drops and reaches the dew point, and water condenses into droplets. A simple analogy is useful to illustrate this relationship: Imagine a bucket which can grow or shrink in response to air temperature, reflective of how warm air has a greater capacity to hold water vapor than cooler air. For this analogy, imagine water vapor as liquid water in the bucket. The total amount of water stays the same, but the bucket size changes. During the day, at 80°F (27°C), the bucket is very large. Depending on how humid the greenhouse is, it may be a little over half full. At night, cold air causes the bucket to shrink down to half its original size. Now the bucket is full and overflowing. Overflowing represents condensation—the point when the air can't hold any more water and it condenses into droplets. Relative humidity is simply the amount of water in the bucket relative to the size of the bucket. In this example, it's roughly 50% during the day, and 100% at night.

Condensation has major implications on material choices, which we touch on in the construction advice in Chapter 9. It means that on most nights over the winter, greenhouses experience droplets of water on all interior surfaces of the greenhouse.

Air Circulation

Outside, wind and convection currents constantly move and circulate air, reducing saturation around the leaf zone. In a greenhouse environment, air circulation must be created with fans, if vents are closed. While not always critical, we find gentle air circulation greatly benefits the growing environment, particularly in the winter.

Air circulation is very easy to create with fans (either wall-mounted or ceiling fans). They do not need to be very powerful: the goal is to move leaves gently, not wildly blow them around. You also want air movement to be uniform throughout the greenhouse (benefiting all the plants). Oscillating wall-mounted fans are helpful to create even air movement in small greenhouses. Larger greenhouses need larger fans, often called Horizontal Airflow fans (HAF). These can be positioned in opposite corners of the greenhouse to make a circular pattern of airflow, or on opposite ends of the greenhouse targeting different areas. In any case, make sure to purchase fans intended for humid greenhouse environments.

Takeaways

- Greenhouse owners typically use electric exhaust fans, or custom-built vents with passive vent openers to ventilate the greenhouse. Exhaust fans provide precise control and quick installation. Passive vents don't use electricity, reducing ongoing energy costs and creating a quiet environment.
- Rule of thumb for sizing passive vents: the total area should equal 15%–20% of the square footage of the greenhouse.
- Rule of thumb for sizing exhaust fans: create at least one exchange of the entire air volume every 1—2 minutes.
- During the winter, close and insulate vents that are not needed for cooling, and add air circulation to reduce the negative effects of high humidity.

Further Reading

Plans and diagrams of vent construction: *The Homeowner's Complete Handbook for Add-on Solar Greenhouses and Sunspaces*. Andrew Shapiro. Rodale Press, 1985.

Endnotes

1. This recommendation has been used by many designers and manuals, including the Department of Energy's evaluation of 200 small-scale solar greenhouses in the 1980s: *Solar Greenhouses and Sunspaces, Lessons Learned*. It is widely used to provide passive venting for solar greenhouses in a range of climates.
2. The vent would need to fit *in between* the studs of the wall, so the actual outer dimensions of the vent would be 14½" or 30½" long if using 2 × 4 construction.

Case Study: Penn and Cord Parmenter*
Penn and Cord's Passive Solar Greenhouses
Residential and community greenhouses
Colorado and surrounding areas

In 1991, Penn and Cord Parmenter arrived on their newly purchased 43 acres in the Wet Mountains of south-central Colorado with a camper van and a wood stove. They were immediately told they would not be able to grow year-round at their 8,000 ft. elevation.

In the ensuing 25 years, they've created an iron forging business, seed business, three kids, and a self-sufficient life revolving around year-round food production.

Through 15 years of practice, they have also become expert solar greenhouse designers/builders. That journey started with a couple of raised beds. Soon after, they built their own small passive solar greenhouse to extend their growing season in the Rocky Mountain climate. Their designs were based on those found in Bill Yanda and Rick Fisher's 1980 book, *The Food and Heat Producing Solar Greenhouse*. In 2006, they started teaching courses on their design and high-altitude gardening at the Denver Botanic Gardens. From there, they began getting inquiries to build the structures for neighbors. In response to the new demand, Cord put down his blacksmith hammer and picked up a claw hammer.

Today, Penn and Cord design and build custom greenhouses in the Colorado region through their business Smart Greenhouses LLC. Their approach is both low-tech and high-impact; their greenhouses are elegantly simple and incredibly robust. As passive structures, they don't include electrical systems or backup heating. Instead, careful planning, thoughtful design and passive thermal mass help to achieve abundant year-round food production, even in very harsh climates. "We've had tomatoes live through −30°F [−34°C] nights," says Penn. Achieving a 60-degree temperature difference without any fossil fuels or electricity is an impressive accomplishment they attribute to "finding the right balance." Heat gain (influenced by orientation, glazing materials and angle) must be balanced with heat loss (determined by insulated wall area, thermal mass and other factors) says Cord.

Penn and Cord are known for using water walls (rows of 55-gallon barrels filled with water and painted black) to give their greenhouses thermal mass and resistance to temperature swings. They are also masters of passive ventilation, which requires careful consideration of wind and snow patterns in the Rocky Mountains. To protect upper exhaust vents, Cord developed a uniquely shaped roof design, shown in Fig. 7.9. It allows vents to be installed vertically on the top of the north wall (protected from snow buildup) while still optimizing the angle of the

---

* Note: Some of the information in this case study was adapted from Bill Giebler's article in *Grit Magazine*: "DIY Principles for a Passive Solar Greenhouse," published January/February 2015.

glazing for maximum light and heat gain in the winter. The south wall typically includes several intake vents. Cord builds awnings over the vents that will stay operable through the winter to prevent them from having to push into snow mounds.

Finally, Cord has developed his own unique method of solving the temperature discrepancy issue associated with solar vent openers. Using a series of connections, he raises the piston to plant level and "transfers the mechanical energy" back down to the opener at the vent. This way, the vent responds to the temperature at the center of greenhouse, not the location of the vent. Penn and Cord Parmenter regularly give classes on their design, building and organic gardening methods. More information is available at pennandcordsgarden.com.

FIGURE 7.9. The north side of one of Penn and Cord's greenhouse builds, also shown on cover, and in the color section. Credit: Smart Greenhouses LLC.

CHAPTER 8

Greenhouse Geometries

Thus far, we've covered the basic principles in greenhouse design—orientation, glazing, insulation and ventilation. The next step is examining how to integrate all these elements into a structure's design. Solar greenhouses can use many shapes and building styles (what we call "a geometry"). There is no single correct design. Rather, a range of geometries can yield the same level of performance and energy efficiency. They vary by their cost, complexity and aesthetics.

Once you determine the basic geometry of the greenhouse, you can move on to creating a framing and construction plan. Thus, this chapter is the bridge between the conceptual principles of solar greenhouse design and the details of construction.

Choosing a Geometry

The shapes of solar greenhouses mostly vary in their roof design, which in turn affects other factors, such as:

- **Material choices:** Designs with vertical or shallow-sloped walls make it easier to use glass as view windows. Some designs allow for one area of glazing, which can be a pro or a con—material and installation costs are reduced somewhat, but two glazing materials allow you to strategically control light by season.

- **Cost and complexity:** The more complicated the geometry of the greenhouse, the more expensive it is to build. Adding angles to the roof design requires more structural supports, which increase the time and cost.
- **Glazing-to-insulation ratio:** We've mentioned before that energy-efficient greenhouses usually use 30%–80% glazing area relative to the insulated wall area. Some geometries create a lower glazing-to-insulation ratio—while still allowing for year-round light—by using a single span of glazing. Other designs use glazing more liberally—on two areas, like vertical walls and the roof.
- **Glazing and wall angles:** Chapter 5 provides a full discussion of the effect of glazing angles on light transmission. Some designs create steeper glazing angles, which will maximize light and heat gain in the winter, though a range of glazing angles are acceptable, as discussed in Chapter 5.

An additional design factor is the angle of the north wall and its role in reflecting light. Some greenhouse designers recommend the north wall also be angled so it reflects light down to the plants. While this sounds valid in theory, the effects of this direct light reflection are minimal unless you are using a wall covering that reflects light in a single direction (called *specular reflection*). Most walls act as diffuse reflectors; they reflect light in many directions. Thus, the angle becomes less important. In our opinion, angling the north wall solely for light reflection is not worth the added time or cost. It can, however, be useful to create geometry such that the south-facing glazing can be a steeper angle, as you'll see in the examples below.

The styles below are the most common geometries for solar greenhouses, and you can see examples of each throughout the case studies in this book. To make a consistent comparison, we use a 12' × 20' greenhouse to depict each design, though they can be scaled to any size. The dimensions we give here are simply examples to put the structure in perspective; they should be modified to suit the particulars of your greenhouse.

A-frame

Consisting of two steeply sloped sides joining at a center peak, this is perhaps the simplest greenhouse geometry, resulting in a cost-effective build.

Pros:

- Simple design; many building plans available.
- A lower-cost design. Using a single glazing material along the entire south wall avoids having to purchase multiple materials and shortens the build time. In the 12' × 20' example, the south glazing area can be covered by 14'-long sheets of polycarbonate.

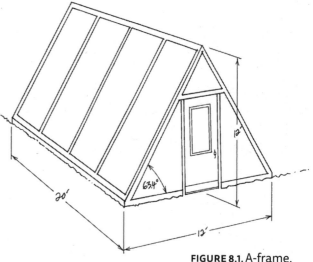

FIGURE 8.1. A-frame.

Cons:

- A-frames can create low headroom at the front and back of the structure. Or, they can create an abnormally tall building if using a smaller width.
- If the entire south wall is glazed (typical), this creates a very large glazing area, reducing efficiency. Glazing near the peak is not effective at transmitting light down to the plants, but it still loses heat.

Shed Roof

One of the most popular greenhouse styles, the shed roof is easy to build due to its vertical walls on all sides. The south wall is usually glazed with glass or rigid plastic. The roof can be partially or completely glazed, and have a range of slopes.

Pros:

- A very common building design most contractors are familiar with. An abundance of plans for sheds or greenhouses is available online.

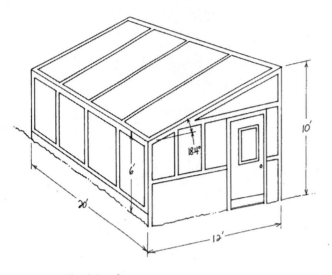

FIGURE 8.2. Shed Roof.

Shed plans can easily be modified using the materials and recommendations in this book.

- One of the simpler and cheaper designs to build.
- It creates two areas of glazing—the roof and the south wall. This has pros and cons in itself: they can be different materials, allowing you to strategically choose materials to tailor light transmission by season (see discussion in Chapter 5), but it could also be costlier to purchase two different materials instead of a larger area of one.
- It enables the use of glass windows because it has vertical walls, creating a nice aesthetic for residential growers.

Cons:

- Both the roof and the south wall have to be glazed in order to achieve adequate year-round light (for both winter and summer months). This creates a larger area of glazing, relative to insulation, which can be a pro or con depending on climate.

Variations: The standard shed design uses a relatively shallow slope to maintain enough headroom on the south side, however a variety of roof pitches can be used to create a steeper roof slope. Though not essential to performance, a steeper pitch is favorable in order to increase winter light slightly (see discussion in Chapter 5) and shed snow in high snow load areas. To create a steeper roof angle, options include:

- Reducing the height of the south wall, turning it into a knee-wall, (shown in Fig. 8.3). This eliminates the need for multiple glazing materials: the knee-wall is insulated and the roof becomes a much longer, steeply sloped plane. A strategic move is to incorporate passive intake vents on the south knee-wall. However, it results in low headroom on the south side.

- Instead of building a vertical wall on the south, this wall is angled slightly, similar to the vertical wall in Fig. 8.6. This adds some building time and cost, since the wall needs a vertical post in order to support the roof, in addition to framing for the angled wall. It also creates a larger footprint of the greenhouse, and creates a larger solar collector for winter light (since the dimensions of the build-ing are greater). Angling the south wall slightly is possible in any of the styles discussed below that have a vertical south wall.

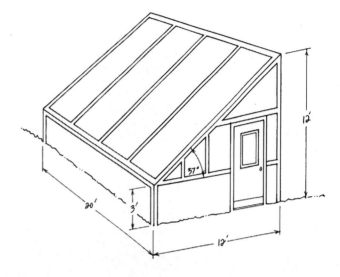

FIGURE 8.3. Shed Roof Variation.

Gable Roof

The gable roof has an on-center peak with vertical walls. The south-facing portion of the roof should be glazed in order to receive sufficient year-round light; the north half of the roof is usually best insulated to reduce heat loss. (Very low-light climates like the Pacific Northwest are an exception in which having both sides of the roof glazed can be justified.)

The gable roof design is a good op-tion if incorporating another structure or use with the greenhouse. The north half of the structure can be a shed area, chicken coop, sitting area, etc. The two areas can be easily divided by a center wall, or kept as one with the north half more shaded for a certain purpose.

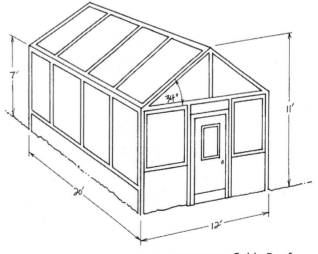

FIGURE 8.4. Gable Roof.

Pros:

- Can be aesthetically integrated with the home, as this is a very common residential style.
- Moderate cost and moderately simple design. Many plans available online for either a greenhouse or shed.
- Many kit greenhouses available online have a gable roof design. You can modify the kit with insulation and proper glazing to make it a more energy-efficient, year-round greenhouse. Keep in mind, though, that the framing of the kit—usually aluminum—probably won't be rated for wind loads/snow loads.

Cons:

- Even with half of the roof glazed, part of the floor area will be shaded during the warmer months when the sun is at a high angle (shading depends on the depth of the structure). Shading is not always a bad thing; it can make the back of the greenhouse a good spot for a sitting area, aquaponic fish tanks, shade-tolerant plants, or thermal mass. Be aware of solar angles during the months you want to grow, and evaluate how much of the floor area will be fully lit using a simple sketch of your greenhouse and the angle of the sun at different times of year.

FIGURE 8.5.
Salt Box Roof.

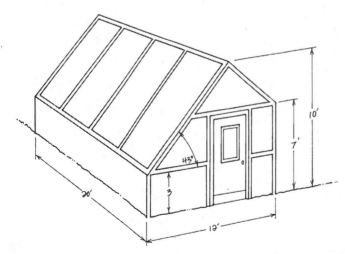

"Salt Box" Roof

This unusual name harks back to colonial America; it was a popular home design named after wooden salt boxes. The salt box roof is similar to the gable roof except that it has an off-center peak. That gives it a larger south-facing roof, which allows for more year-round light, while keeping the overall height of the structure reasonable.

Pros:

- This has a very well-balanced glazing-to-insulation ratio of about 50%–60%.
- The peak gives the south-facing roof a steeper slope, which increases light transmission in the winter and reflects more light in the summer.

Cons:

- This design uses several more angles, which creates a more complex structure to build. The roof and each wall require supports, which can get quite extensive in large greenhouses or those with high snow loads. Large roof spans often require trusses, which further increases building costs in large greenhouses.

Variations: The south-facing roof can be angled to increase light penetration in the winter, shown in Fig. 8.6.

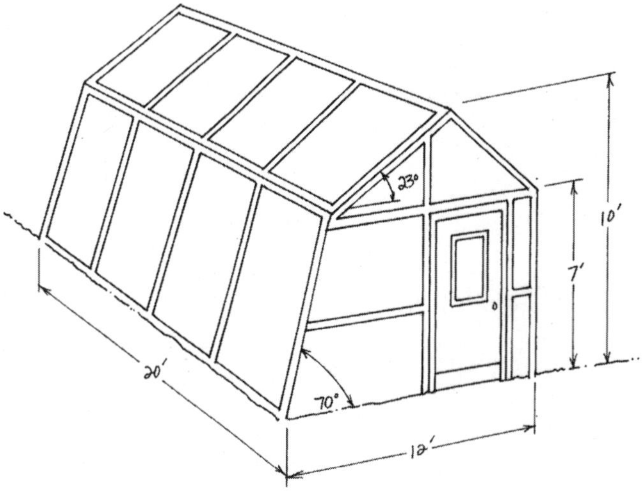

FIGURE 8.6.
Salt Box Roof Variation.

Arched Greenhouses

In the 1980s, Chinese farmers started the trend of building greenhouses with an arched south wall, low insulated north wall, and low-cost materials. As these designs became more popular, eventually gaining some attention in North America, the name came with it, and *Chinese greenhouse* is still how this particular design is best known. Not all arched greenhouses follow Chinese methods, though the two have almost become synonymous in recent years.

The Chinese style of using arched greenhouses (shown in Fig. 8.7) is to use a low north wall (usually built with low-cost natural materials) and polyethylene glazing spread over arched wood beams. These greenhouses commonly include a thermal curtain or insulation blanket (often

FIGURE 8.7.
Chinese-style
Greenhouse. Credit:
indoorgardenhq.com

straw) that rolls down over the south wall at night. This provides a very low-cost method of reducing heat loss through the polyethylene at night. The Chinese design is often used for low-cost commercial structures that are still very energy-efficient.

The basic design of an arched greenhouse can also be applied to more high-tech structures with advanced materials, such as the solar-powered aquaponics greenhouse shown in Fig. 8.8. The pros and cons below apply to the materials and methods used by traditional Chinese greenhouses.

Pros:
- Low-cost and high-performance structure that can be easily scaled.
- Moderate glazing-to-insulation ratio.

Cons:
- If using plastic film as a glazing, this has a short lifespan, particularly in harsh environments. It is hard to find durable insulating glazing materials that can also be easily curved; thus these designs often use thin, low R-value glazing.
- Insulating curtains require a large boom and motor to be automated.

FIGURE 8.8.
Aquaponics Green-
house. Designed
by Friendly Aqua-
ponics, Inc., an arched
greenhouse uses
ETFE for curved
glazing. The year-
round greenhouse
uses plenty of passive
ventilation to grow
through hot summers
in Tennessee. Credit:
Friendly Aquaponics, Inc.

Without the nighttime insulation, the exposed plastic film is very inefficient, negating the overall efficiency of the design.
- The support structure typically involves longer build times (given natural building materials and/or arching beams).

Geodesic Domes

Domes are the most unique of the styles described here. In a dome, small sections of glazing—either pentagons or octagons—are cut out and fitted over the frame (wood or aluminum). They require many more framing pieces; however, the framing members are usually thinner.

There are many true statements online about the dome's unique characteristics—along with a lot of misinformation. Many people cite the dome's increased structural strength as a reason to use it. In truth, any structure can achieve the same strength; it is simply a matter of engineering. What is true is that the domes use thinner framing pieces to achieve the same strength. This is because the framing is spread out over the shape of the dome and many pieces. In a rectangular structure with a flat roof, the structural loads are carried by a few larger beams. That has pros and cons, described below.

In our view, dome greenhouses can perform as well as rectangular structures, but they often require more time and labor to build and insulate due to the many individual sections of glazing or walls. For most growers, deciding whether or not to build a dome usually comes down to the unique shape, which has a certain aesthetic both inside and out. To some it is cool and distinctive; to others it's reminiscent of the 1960s/70s (when this style originated), and doesn't blend with their home/other structures. Often, deciding whether or not to build a dome is simply a matter of personal preference.

Pros:

- Kits are available online which offer prefabricated framing that is easily put together.
- Due to thinner framing members, there is less distinct shading than structures with larger roof beams, but there is not a huge difference in light levels overall.
- Domes are more aerodynamic in high winds because wind blows uniformly over the curved shape. However, high winds can still be problematic for vents protruding from the greenhouse frame.
- As a geometry, the dome has the smallest surface area relative to the footprint. Assuming some of the structure is properly insulated, this can create a very "effective" glazing area: one that lets in a lot of light while reducing the surface area as much as possible.

Cons:

- Most domes, particularly kits, don't include an insulated north wall, which greatly reduces performance in cold climates. To create an energy-efficient structure, the north half should be insulated, but the options with a dome are more limited. Insulating a dome is best done with spray foam insulation, shown in Fig. 8.10. The other option is to cut out individual sections of rigid foam board insulation, which is a much more time-intensive process than it would be for a rectangular greenhouse.

- The increase in framing members creates more areas where air can leak through seams, increasing air infiltration and possibly water leaks if not sealed well.
- Generally, domes take longer to build if doing the framing yourself since there are many more framing sections and individual cuts. Framing-in rectangular openings for fans or doors is more time-consuming.

Takeaways

- The shape of the greenhouse is not a "make-or-break" factor for how it will perform. Rather, performance depends on what you do with that shape—how you insulate it, glaze it, orient it, and control the environment.
- The design does influence material choices, the glazing-to-insulation ratio, and greenhouse dimensions.
- Aesthetics, cost and the complexity of the designs are factors you need to consider early on.

Case Study: Bigelow Brook Farm
A DIY Dome

1,200 sq. ft. residential aquaponics greenhouse
Eastford, Connecticut

Rob Torcellini is a DIYer extraordinaire. An engineer by training, he built his 32' dome greenhouse along with all of its climate and operating systems, documenting his work in a series of high-quality online videos. Systems include a 3 kW solar PV system with battery backup, aquaponics growing system, advanced rocket mass heater, wood stove, and thermostatically controlled vents.

The combination of the solar PV system, rocket mass heater and wood stove allows the greenhouse to run off-grid for about eight months a year. During the low-light winter months, Rob switches the greenhouse to grid power by connecting it to the main breaker panel. Even when connected, the roof-mounted off-grid PV system supplies power for the intricate plumbing of the aquaponics system and other small electric loads. The rocket mass heater and wood stove together provide all heating for the fish tanks and keeping the greenhouse above freezing through the cold and cloudy Connecticut winters. (More details on Rob's modified rocket mass heater appear in Chapter 15.) The aquaponic growing beds are filled with strawberries, vining tomatoes, pine-

FIGURE 8.9. Bigelow Brook Farm Dome Greenhouse. Credit: Bigelow Brook Farm

apples, and an assortment of other crops that form concentric circles within the dome. They are fertilized by the fish living in a 1,000-gallon tank in a large shed area attached to the north side of the dome.

Rob elected for a dome so that he could cut all the parts during the winter and have it ready to assemble in the spring, putting it together like a kit. He also liked that the individual framing members were small enough that he could do the framing himself. The large triangles (spanning 6' across) are framed out of a rot-resistant cedar and sided with polycarbonate. To insulate the north half of the structure, he used spray foam applied in the framing of each triangle. The dome is outfitted with custom vents formed out of single triangle sections. Rob used motorized vent openers to give him more precise temperature control of the ventilation. "Each vent can be programmed to open at different temperatures so they don't have to open all at once, which could create fast temperature swings," says Rob.

To learn more about Rob's design and many systems you can see his many videos on Bigelow Brook Farm's YouTube page: youtube.com/user /web4deb.

FIGURE 8.10. Dome Insulation. Credit: Bigelow Brook Farm

CHAPTER 9

Greenhouse
Construction Basics

The previous chapters should allow you to sketch out your solar greenhouse design on paper. This chapter dives into the principles of building that greenhouse. Our aim is to give an overview of different building methods so you can find which is right for you. With that in mind, we don't provide step-by-step instructions for each building type. We invite you to use the resources listed to explain the details once you find a building method applicable to you.

The construction methods below mostly pertain to building a freestanding structure from scratch. We'll cover prefabricated and "kit" options at the end of the chapter. This distinction raises a common question: should I build the greenhouse myself, hire a contractor or, purchase a kit? The answer depends on your budget and construction experience. Greenhouses are relatively simple structures (much like sheds), and thus most people who have some construction experience feel comfortable building a residential greenhouse themselves. Even if you haven't built anything before, it can be a good project to learn on, though we recommend getting a contractor or experienced builder to help. Larger, commercial structures almost always require an experienced contractor and building crew.

FIGURE 9.1. Table of Construction Methods.

Construction Type	Best for Sizes	Summary	Pros	Cons
Stick Frame	Small to mid	Standard wood construction. Recommended for most residential greenhouses.	• Easy access to building materials • Knowledgeable contractors • Flexible design • DIY • Easy to get permits	• Wood is higher maintenance than metal • Need space to store construction materials • Longer construction times for large greenhouses
Pole Barn	Mid to large	Wood construction method intended for larger greenhouses; Requires building experience.	• Faster installation cost effective for large structures • No need for foundation • Easy to get permits	• Requires a crew of workers • Hard to keep exact dimensions
Aluminum Frames & Kit Greenhouses	Small to large	Metal frames for a range of sizes, from DIY assembled kits to large commercial structures.	• Small kits are easily assembled (fast installation) • Low-maintenance (no repainting)	• Reduced energy-efficiency through the frame • Durability issues with kits • Limited customization
Galvanized Steel Frame	Mid to large	Durable framing for mid–large greenhouses.	• Strongest building material. (Good for areas with very high snow or wind loads) • Some kit options available	• Can be more expensive • Requires professional installation crew • Requires expensive foundation
Structurally Insulated Panels (SIPS)	Small to large	Prefabricated building method using wall sections. Great technology; challenging to find the right products for greenhouse use.	• Quicker installation times • Installers don't need to be experts • Higher insulation levels and overall energy-efficiency	• More expensive • Hard to source panels with water resistant facings • Slightly harder to get permits
Natural Building Methods	Small	Earth-friendly materials such as straw bales and cob. DIY building method that requires extra time and research.	• DIY • Low-cost • Earth-friendly building materials	• Longer building times • Difficult to make them water resistant/potential for mold growth • Hard to get permits

Small = <1,000 sq. ft. / Mid-size = 1,000 to 3,000 sq. ft. / Large = >3,000 sq. ft.

Construction Methods

How you build the greenhouse primarily depends on the size of the structure. Residential solar greenhouses are most commonly built with wood framing; however, stick framing a structure becomes much less economical for larger commercial greenhouses.

Figure 9.1 shows the most common building methods for greenhouses if building a greenhouse from scratch. We've ranked them by the size that they are most appropriate for, since this is a variable you probably already know.

Wood Frame

Conventional wood framing (also called stick framing) is the most common building method (used in almost all homes). Stick framing uses 2×4, 2×6 or 2×8 studs to frame the walls of the greenhouse. Wood can also be used for the rafters to support the glazing, but should be painted or stained if exposed to the humid air.

It is typically the most economical and easy building method for small greenhouses. Materials are readily available, often found used or for cheap. All builders are familiar with this method, making it easy to outsource the labor if not building yourself. Stick framing a small greenhouse can be done by one person, which makes it suitable for a home project if you are building yourself. It's also easy to customize the structure to match your design and needs. For those reasons, we normally recommend stick framing for small to midsize solar greenhouses.

Wood framing requires a framing plan, also called construction drawings. These are detailed drawings that show the exact location of framing members (e.g., studs) and their connections. A full framing plan typically includes multiple drawings, one showing each side of the building (called elevation drawings). It may also include "slice-through" view showing the interior of a wall section.

You can purchase some standard building plans from a greenhouse designer/construction company. Or, you can have a greenhouse designer, architect or builder create a customized one for you. Another option is to modify an existing plan. Quite a few simple ones are available for free online, such as the one from *Mother Earth News,* shown in Fig. 9.2. Books like *How to Build Your Own Greenhouse* by Roger Marshall contain example framing plans. They may not always be the exact size or style you are looking for, but the plans can, of course, be modified. Keep in mind that plans you find online are not usually engineered for specific snow or wind loads. If you live in an area with high snow or wind loads, we strongly recommend getting a structural engineer to review or engineer the plans, ensuring that you're building a durable structure for your climate.

The disadvantage to using wood as a building material is that it does not hold up to moisture as well as metal. Exposed wood requires staining or painting every few years. However, most wood is housed within a wall; typically, only the exposed rafters need maintenance (repainting).

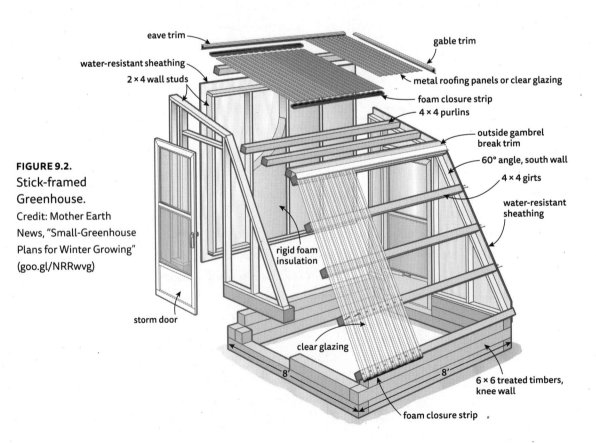

FIGURE 9.2.
Stick-framed Greenhouse.
Credit: Mother Earth News, "Small-Greenhouse Plans for Winter Growing" (goo.gl/NRRwvg)

FIGURE 9.3.
Stick Frame
Construction.
Credit: Ceres
Greenhouse
Solutions

Q&A

If I want a custom greenhouse, do I need an architect or engineer?

This all depends on the size and complexity of the project. Builders of small residential greenhouses normally forgo an architect, as the building plans for greenhouses are very simple. They don't require complicated framing or designs that architects do best. Though someone needs to create a detailed design of the structure, this could be you (if you have construction experience), a builder, or a greenhouse designer. Commercial structures do typically involve architects or professional greenhouse builders who can create construction documents.

Structural engineers review the framing plan and calculate what materials and connections are needed to withstand the structural loads (e.g., wind, snow, and the weight of the roof). He/she will redline the original framing plan, calling out the exact materials and connections. The decision to seek an engineer's review depends on the size of the build, who created the plan, the wind and snow load at your site, and whether the greenhouse needs a building permit. Builders of unpermitted residential greenhouses typically skip the structural engineering, relying on their experience and knowledge to build a durable structure. Larger greenhouses, and those that need a building permit, require engineering. Building plans can also be pre-engineered. For example, at Ceres, we have sets of building plans that have been engineered for a set wind and snow load and now can be used in a variety of situations without the guess work.

The residential structures we have built mostly used standard lumber for the framing, for both walls and rafters.

The primary disadvantage of stick framing a greenhouse on-site is that it is typically time intensive. Whether built by you or a by a contractor, your backyard may become a construction zone for a few weeks. If this dissuades you, there is an additional option of having the structure prefabricated by a greenhouse or shed builder.

Prefabricated structures are constructed in a shop and then delivered as wall sections (or a full structure) to the site, as shown in Fig. 9.4. The building process takes a couple of days, compared to weeks of on-site work. The downside is that prefabricated structures are typically more expensive than building on-site once shipping/transportation is taken into account. It can also be challenging to find a local builder who offers this option. For those reasons, standard on-site framing is more common for residential and mid-size greenhouses; it is often a do-it-yourself project, or it can be hired out from a local contractor.

Pole Barn

Pole barns, also called post and beam structures, are a simplified building style used for larger structures. They involve burying lumber posts deep in the ground, or setting them in concrete piers. The posts give the structure stability. Instead of a load-bearing *wall* comprised of many

Choosing a Contractor

If hiring out the construction work from a contractor, in a perfect world he or she would have experience building greenhouses. Since this is often not the case, you may have to inform the builder about some the differences in greenhouse design and construction. For example, glazing materials and planning for a humid environment are not standard procedures builders are familiar with. While most builders can easily build a small to mid-size greenhouse, you may have to alert them of these variations using the designs and recommendations of this book or an experienced greenhouse designer. Beyond that, standard good-sense practices apply when choosing a contractor for the project (e.g., check their references and liability insurance).

FIGURE 9.4.
Prefabricated Walls.
Credit: Ceres Greenhouse
Solutions

2 × 4s, pole barns have load-bearing *posts* spaced several feet apart. The posts can be made out of treated or laminated lumber. Horizontal boards (called girts) connect the posts to form the wall.

As the name implies, this construction type is commonly used in barns and other large simple structures. For large greenhouses, it is typically cheaper than a conventionally framed structure because it requires less framing. It also simultaneously provides the foundation: the posts are buried directly into the ground and supported with minimal concrete footers or piers.

A downside to pole barns is that they require an experienced and accurate builder. It is difficult to get the spacing of the posts exactly right, since they are set in place before there is any supporting structure (e.g., base frame). If distance between posts is off, the rest of the materials (roof panels, insulation panels, etc.) will be off and need to be adjusted. Additionally, like standard construction, exposed wood needs protection from humidity. We find pole barns less common as a building type for larger greenhouses. Metal buildings, such as aluminum and steel are more common as a durable and cost-effective construction type for large commercial greenhouses.

Aluminum Frames and Kit Greenhouses

Aluminum is a popular building material for small and large greenhouses alike. Search "kit greenhouse" online, and most of the options you first see are either PVC or aluminum frames. Metal frames require less maintenance and last longer than wood structures. The disadvantage is that aluminum is an excellent heat conductor. Aluminum quickly leaks heat to the outside.

Choosing the Right Kit

Typically, aluminum frame greenhouses come in the form of "kits"— structures with pre-cut parts that can be easily put together. This is an enticing option for many growers who are not inclined toward building. However, it's one that necessitates some caution. Kit greenhouses are not rated for snow or wind loads. They are designed as low-cost, "one-size-fits-all" structures, intended to be cheaply shipped around the country. They are also not designed with energy efficiency in mind. The greenhouse kits you will find on the top of a Google search will generally not be year-round solar greenhouses; they are likely to be flimsy and inefficient structures.

Standard greenhouse kits can be handy as season-extenders in many areas (again, keep snow and wind in mind). Due to their many drawbacks, we generally don't recommend them for year-round growing in most climates.

Another option is to purchase a better-quality kit and modify it to be more energy efficient. This involves adding rigid foam insulation in sections between the frame and reducing air infiltration wherever possible (using caulk or an insulating spray foam). Some companies may allow you to customize the kit to rate it for high wind and snow loads. We recommend paying close attention to the warranty that is offered, the quality of the glazing materials (is there one layer or two?), and reviews about the product. Since this is time intensive and provides a

FIGURE 9.5.
Blown-over
Aluminum
Frame.

lower-quality structure than a self-built greenhouse, we believe you'd be better off spending the time and money on building a custom solar greenhouse from the start.

Galvanized Steel Frames

Steel is the strongest framing material, and thus often used in large commercial greenhouses. Steel structures can be insulated in a variety of ways. At Ceres, we combine them with insulated metal panels, commonly used in warehouses or large industrial buildings. These are similar to structurally insulated panels (SIPs), discussed below, but have a metal facing. Due to the possibility of rust, steel frames must be galvanized. They are cost-competitive with aluminum frames.

Structurally Insulated Panels

Structurally insulated panels, or SIPs, are a unique construction type that holds a lot of potential for greenhouses, but some challenges as

FIGURE 9.6.
Steel
Construction.

well. SIP structures are made out of prefabricated wall sections. The panels are usually made of an insulated foam core sandwiched between two facings, typically oriented strand board (a wood composite known as OSB). The foam insulation is blown into the cavity in the factory, and the structure is delivered to the site in panels, usually 4' wide. The sections are then put together a lot like Lego pieces, as shown in Fig. 9.7, and then bolted/screwed together.

SIPs are an entirely different construction style. In contrast to stick framing, the panels themselves provide the structural stability, and there is little framing in the wall. For that reason, SIPS create a more insulated wall overall, creating a continuous layer of thick foam insulation. SIPs have very high insulation ratings, ranging from R-21 to R-52.

Another advantage is faster installation times. Since the walls are prefabricated at a factory, a residential greenhouse can be assembled in as little as a day with a small crew of three or four people. (Panels are very heavy, so at least two people are required to assemble the structure.) The methods of attaching the panels to one another vary by product, but all are relatively easy to complete by a handy homeowner with some helpers, provided proper instructions.

SIPs may seem like the perfect building material—prefabricated, easily assembled, highly insulated—but, there are a few major hurdles given current manufacturing. The standard facing material is OSB,

FIGURE 9.7.
SIPs Construction.

which performs poorly in humid environments and shouldn't be used in a greenhouse. Thus, we recommend SIPS with a different facings. There are a couple of manufacturers that use a water-resistant facing called magnesium oxide (MgO). MgO SIPs often need to be shipped from far away since sourcing is limited (see Further Reading for manufacturers). Furthermore, MgO is quite brittle, and corners can easily break off the panels.

SIPs are about 20%–30% more expensive than conventional building methods, though much of these costs are saved by shorter construction times. If manufacturing becomes more common, we see SIPs with water-resistant facings as an excellent option for energy-efficient greenhouses in a wide range of sizes.

Natural and Recycled Materials

Natural building materials—such as straw bales or cob—are also possibilities for residential greenhouses, though less common. While these are good options for sustainable building methods in homes, greenhouses add the challenge of constant humidity, which can lead to mold growth in natural materials. There have been a number of successful projects that involve cob or straw bales (see Fig. 17.7 and "Strawbale Greenhouse" in the color section for examples). However, we don't recommend these due to the high risk of mold growth that results from even the smallest errors in installation or plastering.

Whether using straw bales or cob, a quality plaster must be applied. Most builders recommend lime-based plasters for a greenhouse application (instead of clay-based plasters) because they are more resistant to moisture. Plastering is not simply a paint job; it should be done by a professional or someone with experience. For more on building with cob or straw bales, we recommend the resources listed in Further Reading, and talking with experienced builders.

The advantages of natural building materials include lower costs and more sustainable methods. They can create a thick, well-insulated wall. Straw bales, for example, can provide an R-20 to R-40 rating depending on the thickness of the bale. Natural building materials can also add to the thermal mass of the greenhouse, as they are often dense materials with higher heat capacities. While cheap in terms of dollars, they do require much more building time, as well as the time it takes to collect/

FIGURE 9.8.
Tire Barrel Building. Called "The Greenhouse of the Future," a community greenhouse project located north of Montreal is made out of recycled tires filled with sand. You can find a DVD course to teach you how to build one at greenhouseofthe future.com

make materials. Often, these greenhouses are small to midsize and are built by a crew of volunteers or community members.

Like natural building materials, using recycled building materials is a great way to cut costs, though it should be done thoughtfully. Carefully evaluate the condition of what you are buying. Double-pane glass windows or storm doors in good condition can make excellent recycled components of a greenhouse. These are often found at resource yards in a range of sizes. Realize, though, that there is probably a reason they have been given away. If you find double-pane windows, carefully check that the frame is in good shape. If the seals are broken, it will cause the window to fog up, reducing light transmission. Single-pane glass is also commonly found reused. This has a very low R-value, though, and it's not something we recommend for good performance except in mild climates. If cost-savings are essential to you, it may be a good choice, though keep in mind how it could reduce the efficiency of your greenhouse.

Foundation Types

If the framing is the bones of the structure, the foundation is the backbone: it stabilizes the structure above ground, preventing the greenhouse from shifting over time and under weather.

The first question is whether the greenhouse needs a foundation. Small greenhouses (under 200 sq. ft.) are often stable enough that a foundation isn't absolutely necessary. In these cases, the greenhouse should be installed on a level base, either gravel or dirt. An additional step is to anchor the greenhouse to the ground using rebar or products called earth anchors to prevent it from lifting in high winds. We recommend these for any high wind areas. If unsure whether your greenhouse should have a foundation, check with your local building codes and/or a structural engineer.

Most solar greenhouses need a foundation for structural stability. Foundation options mirror those for homes or commercial structures. Factors to consider include permitting requirements, cost and the flooring in your greenhouse.

Concrete Piers

Piers are tubes of concrete buried in the soil to a depth below the frost line, typically 3'–4' below grade. They are installed by first digging a hole in the ground. A plastic or fiber tube form is then set into the hole and filled with concrete. While still wet, anchor bolts are set into the pier. These attach to the base frame of the greenhouse. A picture of concrete piers, combined with underground insulation is shown on the last page of the color section. They're spaced roughly 4'–6' apart, depending on the loads of the structure. They are a moderate cost, much faster to install compared to the methods below. The forms and concrete are readily available at hardware stores. For those reasons, we recommend a concrete pier foundation for most greenhouses. They are suitable for a wide range of sizes and applications.

Concrete Wall and Footer

This is the typical foundation for houses. It consists of a concrete wall that extends below the frost line, anchored by a footer. A concrete wall foundation is a very expensive option, but often necessary for attached greenhouses. (Permitting requirements usually stipulate that an attached structure should match the foundation of the house.) It's also useful for greenhouses built into slopes where a stem wall (concrete retaining wall) is needed, as shown in the SIPS greenhouse in Fig. 11.5.

Slab on Grade

A slab on grade foundation is a slab of concrete a few inches thick, with a thicker footer around the edges. The slab forms both the foundation and floor of the greenhouse. This method should only be used if you are not planning on growing directly in the soil underground or in raised beds with access to the soil below grade. A concrete floor blocks access for the roots to penetrate the soil. However, it is useful for other growing methods that may need a level floor for heavy or rolling equipment. It's commonly used in hydroponic or aquaponic greenhouses, as well as commercial structures that need a floor that can be easily cleaned. The slab should have a floor drain, as shown in Fig. 9.9.

FIGURE 9.9.
Concrete Slab and
Footer Foundation.
Credit: Ceres Greenhouse
Solutions

Insulated Concrete Forms

Insulated Concrete Forms, or ICFs, are interlocking blocks of foam insulation (often polystyrene, but insulation types vary) with a central void. The blocks are stacked on-site, braced, and then filled with poured concrete. The concrete is thus sandwiched by two layers of foam insulation. In this way, ICFs provide both the foundation wall and underground perimeter insulation, two benefits in one system. ICFs create highly insulated foundations, with R-values around R-20.

ICFs are also commonly used in walls as well, though we don't recommend them for above-ground construction because they are difficult to customize. Any openings in the walls, like windows or vents, must be cut through concrete. They do have applications for underground walls—in an underground or earth-bermed greenhouse—as exemplified by the case study at the end of the chapter.

ICF foundations are often faster to install than poured concrete walls, because instead of building forms on-site, the prefabricated blocks create the form of the foundation wall. However, they still require care in installation and, ideally, a professional with experience. Errors in the construction process, like a block shifting, can be extremely costly.

Costs for ICF foundations are in the same range as poured concrete walls. In short, they are excellent choices from a performance perspective, though a relatively expensive option.

Siding and Finishes

Decisions about the right sidings, caulking and vapor barriers rest primarily on one all-important characteristic: greenhouses are humid. Recall from Chapter 7 that this fact of life results from plants' ongoing respiration of water vapor into the air. Due to the humidity and temperature fluctuations, greenhouses regularly experience condensation at night. Condensation in a greenhouse occurs most nights in the winter; it is less common in warmer months because the greenhouse is ventilated.

Condensation makes the inside of a greenhouse more like an outdoor environment than an indoor one. This requires careful choices of siding materials. They should be water resistant or water impermeable. We recommend simply using those intended for exterior applications.

Do NOT Use These Sidings:

- **Drywall or Oriented Strand Board (OSB):** These are the worst possible choices. Water can easily penetrate the material, causing it to swell, degrade and mold.
- **Standard plywood:** Standard plywood is better than OSB, but water can still penetrate between the glued-together layers of board. Most grades are not rated for exterior use, and the glue is not water resistant. Pressure-treated types have more resistance to rotting, but introduce toxic chemicals into the greenhouse that can leach into the soil. If considering plywood, it should be regularly painted.

Good Siding Options

Because untreated woods don't hold up well to moisture, and pressure-treated woods contain chemicals, we recommend using non-wood alternatives for siding. These include:

- **Fiber cement/composite sidings:** Composite siding materials are made out of wood and fiber cement products. Brand names like HardiePanel or LP SmartSide are water resistant and good choices for interior siding. Both come in sheets, normally 4' × 8' that can be easily installed in stick frame greenhouses.

- **Magnesium oxide (MgO) board:** Magnesium oxide is a naturally occurring mineral. As a siding material, it comes in sheets (4' × 8' or 4' × 12'). We often describe it as a "stone drywall." It's made out of a mineral, so it doesn't rot or degrade with moisture, and it's is also fireproof. A downside is that it's brittle. Corners can easily snap off or cracks can form that need to be repaired. Secondly, it must be sourced from specialty distributors. It is cost-competitive per square foot, but can get quite pricey if shipped long distances.

- **Fiberglass and metal sheets:** There are several types of plastic panels available; it's the material you would normally find in a bathroom

Should you use pressure-treated wood in the greenhouse?

Prior to 2003, pressure-treated wood was injected with a chemical mixture that included arsenic, a well-known poison. Called CCA, the treatment was banned in 2003 by the US Environmental Protection Agency. In CCA's stead, an array of other chemical treatments has emerged, some more benign than others. On the more benign side are those that use copper as the active ingredient, such as ACQ or MCQ, but do not contain arsenic or other chemicals considered toxic by the EPA. Borate-treated lumber adds boric acid to wood, which protects against rot and insects but is considered very safe for humans.

Though it varies by individual product, the overall consensus among scientific studies is that use of pressure-treated woods in garden beds does not pose a significant threat to plants or human health. Though the chemicals do leach into the soil, plants only take up chemicals in very small amounts. It is highly unlikely that these accumulate enough in plants or in people to pose a significant threat to either.[1]

Though the risks are small, we recommend erring on the side of safety and only using pressure-treated woods when absolutely necessary. Their effect on soil biology is less known, and these are certainly not organic compounds. We commonly use pressure-treated wood for the base frame of the greenhouse, but restrict its use elsewhere in the greenhouse, where we use standard wood or composites instead.

shower. A variety of options in fiberglass or acrylic are available in sheets at hardware stores. Metal is another option. Aluminum sheets are widely available, but usually pretty expensive.

- **Cedar:** Given its resistance to moisture and rod, cedar lap siding (or other durable, long-lasting wood) is a good option as well.
- **Marine-grade plywood:** Some types of plywood are rated for exterior use, or even boat applications. Though more expensive, this may be a good choice if you find an economical or local source. We don't have enough experience with its durability to comment resoundingly; some greenhouse designers recommend it.

Caulk and Vapor Barriers

Caulking is an often overlooked, but critical, part of finishing the greenhouse. It's imperative to keep the inside of the wall cavity dry. Not only can moisture lead to mold and mildew, but when insulation gets wet, it loses much of its insulating ability. For that reason, we caulk all seams and screw holes to create a "tight" structure and prevent moisture penetration into the wall.

While caulking is mandatory, vapor barriers are an optional addition that can be used for extra protection against mold in the wall. An option is to install a semi-permeable vapor barrier on the interior side of the wall. Some solar greenhouse designers strongly recommend vapor barriers to help keep the wall cavity dry; we normally forgo them and rely on caulk and quality sidings to prevent water penetration into the wall.

Some final tips to plan for humidity:

- Bevel all sill plates underneath the windows so that water can easily drip off, as shown in Fig. 9.10. This is an area where water accumulates as condensation drips down the windows at night. A slightly angled sill reduces water sitting on a sill and thus damage to the material.
- Use stainless steel or galvanized screws on the interior (also called deck screws). Standard screws will rust. The rust runs down the wall, making the walls of your green oasis very unsightly.
- As described more in Chapter 16, all outlets should have covers.

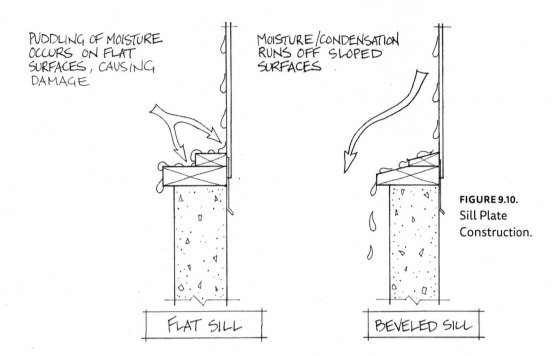

PUDDLING OF MOISTURE OCCURS ON FLAT SURFACES, CAUSING DAMAGE

MOISTURE/CONDENSATION RUNS OFF SLOPED SURFACES

FLAT SILL

BEVELED SILL

FIGURE 9.10.
Sill Plate
Construction.

Takeaways

- For residential and mid-sized greenhouses, stick frame construction is most economical and easiest. Other types of construction, such as using structurally insulated panels, pole barn framing, or natural building materials are viable, but take more research and experience. Metal buildings are most economical for larger commercial greenhouses.

- If opting for a kit greenhouse, it's important to choose a quality kit that will withstand wind and snow loads at your site. Often, kit greenhouses are not insulated, so they usually have to be retrofitted to be more energy efficient. Check warranties and the quality of the specific materials.

- Almost all greenhouses require a foundation. Concrete piers are the most cost-effective type.

- Choose sidings and finishes carefully; consider the effects of a high-humidity environment.

Further Reading/Resources

Stick Frame Construction Instructions and Building Plans

Marshall, Roger. *How to Build Your Own Greenhouse*. Storey Publishing, 2006.

Freeman, Mark. *Building Your Own Greenhouse*. Stackpole Books, 1997.

Kits and Building Materials

Ceres Greenhouse Solutions provides insulated greenhouses in wood and metal kits (options vary by location and sizes): ceresgs.com

Sturdi-built Greenhouses is based in Portland and provides cedar kits: sturdi-built.com

Growing Spaces is based in Colorado and provides geodesic dome kits: growingspaces.com

Article on choosing a prefabricated greenhouse kit: "How to Choose the Best Greenhouse Kit," Shane Smith. *Mother Earth News*, goo.gl/ekOXWK

MgO SIPs: Companies currently manufacturing these panels include MAGpro Building Systems (mag-pro.com) or MegCreteE (rccorp.com), both based in Alberta, Canada.

Endnotes

1. Richard Stehouwer, "Environmental Soil Issues: Garden Use of Treated Lumber," University of Pennsylvania, extension.psu.edu/plants/crops/esi/treated-lumber

2. roperld.com/science/ymcasolargreenhouse.htm

Case Study: YMCA/Roper Solar Greenhouse
Building a Greenhouse Step-by-step

576 sq. ft. community greenhouse
Blacksburg, Virginia

In addition to typical community programs, the YMCA at Virginia Tech offers members a unique opportunity: a winter gardening space in their earth-bermed solar greenhouse. Designed by L. David Roper, a retired physics professor, and architect Tim Colley, the community greenhouse was built by volunteers and community members over the course of several months. It now hosts frequent educational tours, as well as year-round gardening plots for community members.

The greenhouse is an earth-bermed design, with an A-frame shape on top. The aboveground walls are wood framed with 2 × 6 or 2 × 10 construction. The south-facing wall is glazed with polycarbonate. The north, east and west walls are highly insulated with 3" of closed cell blown-in foam. Below ground are a few feet of vertical walls constructed out of insulated concrete forms.

The greenhouse is heated and cooled during winters with a ground to air heat exchanger, described more fully in Chapter 12. It is ventilated with awning windows, operated with thermostatically-controlled motorized window openers. An exhaust fan provides back-up venting. Finally, an outdoor underground cistern collects rainwater from the roof to provide all watering for the greenhouse.

Mr. Roper documented each step of the building process, making them available via his website. Though the online resource does not show every swing of the hammer, or offer plans for building your own, it does provide an excellent tool to understand the steps involved in building a community greenhouse. A meticulous documenter, Mr. Roper includes a timeline of each activity and milestone, so viewers can get a feel for the time required to go from plans to finished structure. For more information, see vtymca.org and roperld.com/science/ymca solargreenhouse.htm

FIGURE 9.11. YMCA/Roper Greenhouse. Credit: L. David Roper

CHAPTER 10

Attached Greenhouses

Attached greenhouses are dual-purpose structures. Not only are they a lush year-round garden, but the greenhouse also serves as a vital heat source for the home. Greenhouses are such large solar collectors, they have the ability to dramatically reduce a home's winter heating bill— 30%–50% in many cases. Even if supplemental heating is not the primary purpose of the greenhouse, attached greenhouses have several benefits over freestanding ones. But, they create some unique challenges as well.

Integrating the Greenhouse and the Home

Pros and Cons

If done properly, the typical attached greenhouse presents many benefits for both the home and greenhouse:

- **Free heat:** A greenhouse is a large solar collector, a source of free home heat during cold winter days. The home, in turn, absorbs excess heat during the day. Combined, the two environments work synergistically, requiring less heating and cooling as one.
- **Supplemental CO_2:** The home provides a source of CO_2 and air circulation for the greenhouse. Plants naturally filter the air and add oxygen to the home.

- **Controlled humidity:** In the winter, the greenhouse adds humidity to the home's drier air. Air from the home lowers the relative humidity in the greenhouse.
- **Added home value:** A site next to the home takes up minimal yard space. The greenhouse can blend aesthetically with the home, adding significantly to a property's value.

On the other hand, attached greenhouses come with some particular challenges:

- **Permit costs:** Most people will tell you an attached greenhouse is cheaper to build than a freestanding one. That is true only if a building permit isn't required. In many places, local building codes consider an attached greenhouse an addition, like an extra bedroom, and mandate the same foundation type as the house (usually a concrete wall and footer) to prevent the greenhouse from shifting and damaging the home. Normally, the project requires stamped engineered construction documents, and a more robust foundation. These can easily add a few thousand dollars to the cost of the project. Even if a permit is not required, attached structures require more careful building, which often means more expense, not less.
- **Excess heat and humidity:** A hot, humid environment adjacent to the home is beneficial for much of the year. At other times, heat and humidity are unwanted and create a climate burden for the home. Excess humidity is probably the number one problem with attached greenhouses: it can damage books, furniture and electronics, or allow mold to grow in the shared wall. For that reason, much of this chapter focuses on controlling air exchange between the two environments so the greenhouse does not over-humidify the home.
- **Pest risk:** If pests, like aphids, get established in the greenhouse, there is the risk that they will spread into the home. Most pests stay where their food source is (the greenhouse), but it is possible for them to expand into the home. Here too, controlled air exchange helps keep pests out of the home. Partitions also help.

Approaches

There are a few different approaches to attaching a greenhouse; which you choose depends on how integrated you want the greenhouse and home to be. At the most basic level, a greenhouse can be *structurally attached* to the home, but not exchange air with it. An open south-facing back of a garage, for example, is an excellent spot for a greenhouse, as shown in Fig. 10.1. Since the site is not adjacent to living areas, homeowners usually don't go through the expense of venting the greenhouse's heat into the home (though it's possible with ductwork). The greenhouse uses the existing wall as its north wall, but otherwise operates exactly like a freestanding structure.

On a side note, there's also the possibility of what we call the "almost attached greenhouse." To get around prohibitive and expensive permitting requirements, some homeowners site the greenhouse about 2' off the house, and make it *look* attached. Technically considered a detached structure according to building codes, a permit is not required, but the

FIGURE 10.1.
Attached Greenhouse.
CREDIT: Ceres Greenhouse Solutions

greenhouse still blends with the home and makes use of a good site that doesn't take up extra space in the yard. A false wall can be built in between the two to block heat-sapping winds blowing behind the greenhouse. (See colour section for an example of an almost-attached greenhouse.) If considering this route, check your local building regulations, which may stipulate how far an accessory structure must be from the house. An example is of an "almost attached" greenhouse appears in the color photo section.

Most attached greenhouses are sited adjacent to a living area, connecting the two environments to take advantage of the free heat of the greenhouse. Because this the most common arrangement, we'll call it the "typical attached solar greenhouse."

The typical attached solar greenhouse is designed mostly like a freestanding one. The major difference is that the greenhouse also exchanges air with the home. The greenhouse should still be equipped with standard ventilation to exhaust excess heat outside when you don't want it in the home. So there are two venting routes: one connects to the home (air exchange) and one to the outside (ventilation). All the other aspects of solar greenhouse design—thermal storage, insulation, careful glazing, etc.—still apply, so the greenhouse can grow year-round without depending on backup heat from the home. The typical attached solar greenhouse is usually retrofitted onto the home and separated, except for some openings in the shared wall.

Another approach is becoming popular with new homes: *fully integrating the greenhouse and the home*. In cases of new construction, the design options abound. Instead of simply an add-on structure, the greenhouse can be a lush, sun-filled space seamlessly blended with the living areas. It can be integrated with a kitchen, or it can occupy the center of the home as its focal point and main heat source, as shown in the case study at the end of this chapter.

While this strategy has great potential benefits, it calls for caution. The greenhouse can easily "over-glaze" the home, making it susceptible to huge temperature swings. While large south-facing windows are a premise of passive solar homes, a greenhouse often requires more glazing than is normally recommended (i.e., passive solar home designers recommend the glazing area equal 6%–10% of the home's footprint; greenhouses create much more). It's essential that there be sufficient thermal mass and enough insulation to accommodate for the glazing. This type of arrangement also is best with more advanced glazing materials to control heat loss. Excess humidity is a potential problem. Installing a heat recovery ventilator (HRV), described in Chapter 7, is a good option. If you are considering building a fully integrated greenhouse, discuss it with your architect or home designer early on in the process, and consult the resources on passive solar home design at the end of this chapter.

Siting

Siting an attached greenhouse follows the principles discussed in Chapter 3. In addition, we recommend siting the greenhouse around existing openings in the wall (windows or doors) to take advantage of passive air exchange between the two. Moreover, doors and windows give easy access to a lush year-round garden just steps away. You get the food and heat of the greenhouse, and the verdant view as well.

In an ideal world, an attached greenhouse would be built slightly below ground. This allows the peak of the greenhouse to cover an existing opening in the shared wall, so that the hottest air in the greenhouse will vent directly into the room. If the two structures are on the same level, hot air rises to the top of the greenhouse and is vented into the home near the ceiling. The hot air stagnates there, doing little for comfort. Thus, a strategic integration is combining an attached greenhouse with an earth-sheltered design (discussed in Chapter 11), with the floor of the greenhouse sitting a couple feet below grade, as shown in Fig. 10.2. The top of the greenhouse can then be located around existing windows in the home, venting warm air directly into the room. This also allows the greenhouse to sit more easily underneath the home's existing soffits or overhangs. It does require more caution in building, since the greenhouse will sit next to the foundation of the home. Consult a structural engineer if considering this route.

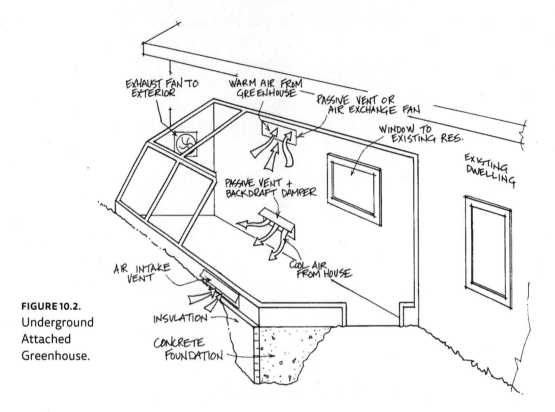

FIGURE 10.2.
Underground
Attached
Greenhouse.

Controlling Air Exchange

Exchanging air between the greenhouse and the home behooves both
structures, but must be *controlled*. Otherwise, the greenhouse can easily
overheat and over-humidify the home. It can also overcool the home at
night—a large uninsulated glazing area will lose a huge amount of heat
during winter nights, potentially increasing heating bills.

Separating the two environments is logical because plants and people have different ideal temperature ranges. Plants have a greater tolerance for cooler temperatures (most down to 40°F [4°C]), whereas
people like to keep their homes above 65°F (18°C). Since the two have
different requirements, trying to heat and cool them as one results in
wasteful overheating of the greenhouse. It is better to equip the greenhouse with sufficient thermal storage and insulation, separate the two
environments and keep the heat in the home at night. The home can

provide a source of backup heat on very cold nights, but we don't recommend relying on it extensively.

The methods for controlling air flow between the home and greenhouse mirror the options for ventilation discussed in Chapter 7: windows and doors, passive vents, and fans. Refer to Fig. 7.1 for a summary of these. Below, we'll discuss how each applies to an attached greenhouse.

Passive Air Exchange

If the greenhouse is sited around windows/doors in the shared wall, manually opening these is an excellent option for venting excess heat into the home. The drawback is that you must be there to facilitate it—usually twice a day—and that makes control less precise and reliable.

If there aren't any existing openings in the shared wall, it is common to retrofit a sliding glass door into the wall. While a considerable added expense, it is often worth it to get the views and easy access to your year-round garden.

Vents and Backdraft Dampers

If there are no existing openings, and you don't want to go through a full retrofit of your wall, an alternative is to install passive vents between the greenhouse and home. These are smaller openings that rely on natural convection cycles to facilitate air movement similar to the passive exhaust vents discussed in Chapter 7. Typically, vents are sized to fit between the wall studs so the framing doesn't need to be altered. Intake vents should be located low on the shared wall, and exhaust vents high. When doing a retrofit like this, consider the home's electrical wiring, plumbing and baseboards before knocking a hole in your wall.

Typically, vents include a simple control mechanism to stop the convection cycle at night. Ideally, you only want air to flow from the home to the greenhouse during the day, when the greenhouse is able to supplement heat. At night, vents should be closed to prevent over cooling the home. Backdraft dampers, either purchased or handcrafted, can be used for this purpose. In older solar greenhouse books, you'll see instructions for making your own using plastic flaps over the vent opening. The plastic flap is installed on the home side of the upper vent and greenhouse side of the lower vent. This arrangement allows the flaps to open the greenhouse is in "heating mode" and stay shut when there is no convection cycle.

Backdraft dampers are cheap (Bill Yanda describes them as the 3-cent solution) and not sophisticated devices. They don't provide an airtight seal or insulation between the two environments. They also don't provide any control against overheating or over-humidifying the home, which are important considerations. For these reasons, we don't recommend them as a means to control airflow. In our opinion, your home and its contents are too valuable to risk damaging from excess humidty. Other reputable greenhouse designers recommend them as a cheap and passive venting method. See Further Reading for more on their techniques.

Active Air Exchange

The third option for controlling airflow is thermostatically controlled fans. The same exhaust fans and intake shutters discussed in Chapter 7 can be used in a shared wall. These allow for precise control of air and heat exchange, but come with the drawback that they are noisy, and can be annoying if located in common living areas. To reduce noise, purchase a fan with a speed controller and set the fan at a slower speed.

Fans should be coupled with intake vents to allow return airflow into the greenhouse, and controlled with a thermostat inside the greenhouse. We recommend going one step further and adding a second thermostat and/or a humidistat inside the home. This requires more advanced wiring, but it's the only way to add regulation based on the *home's* temperature and humidity level. In this scenario, a thermostat in the home turns the fan *off* when the home reaches a critical temperature. Thus, the fan can only run when two conditions are met: an over-heated greenhouse and under-heated home. You can also purchase a thermostat combined with a humidistat to control for home humidity levels as well.

In our view this strategy is the safest way of preventing overheating and over-humidifying the home, and it's what we recommend to control air flow (wisely combined with manually opening doors, when you are at home). It requires a more advanced wiring layout that should be installed by an electrician if you do not have electrical experience.

Combining Ventilation (Outdoor Air) and Air Exchange

As we've mentioned, the typical attached greenhouse needs two ventilation routes—one connected with the outside air and one with the home. It's important that these work congruously. The goal is to have air exchange vents open *first*—venting excess heat into the home—and *then* have vents to the outside open if the greenhouse needs additional venting. This is where thermostatically controlled openers and precise control comes in handy. If orchestrating this scenario with passive vents or solar vent openers, there's a good chance vents will open at the wrong time.

The balance between air exchange and standard ventilation varies by season. For most climates, in the summer all air exchange vents or fans can be closed, because the home will probably not need heating. During the winter, many of the exterior exhaust fans or vents can be closed and sealed as described in Chapter 7, since the majority of greenhouse's heat will be used for the home.

Controlling Heat and Humidity

In addition to limiting air exchange in the summer months, we recommend adding a shade cloth to the greenhouse if you live in a warm climate. Basically, you want to take every measure to reduce heat gain that will overheat the home. Even when the two areas are separated by a door or window, direct light shining into the home can add uncomfortable heat.

Roof insulation can be strategically placed to shade windows and doors, but still allow light into the growing area. An insulated roof is used like an overhang, a common feature in passive solar homes. See the section in Chapter 12, titled "Water as Thermal Mass" for more on calculating the length of roof insulation.

A minor issue with attached greenhouses and sunspaces is that air can stagnate in one room, creating a single hot room without much benefit for the rest of the home. This can be compensated for with a series of fans to move air into adjacent rooms; luckily, even after the greenhouse is constructed, this is a relatively easy remedy to apply.

We also recommend additional measures to monitor and control humidity in the home. According to Andrew Shapiro, author of *Add-On Solar Greenhouses & Sunspaces*, "a residential greenhouse full of plants can move as much as 2 to 3 gallons of water into the house over the course of the day." This moisture is mostly an issue for the surfaces directly exposed to the greenhouse: window and doorframes and the shared wall between the greenhouse and home. It can also create condensation on windows at night.

Whether humidity is an issue depends on the size of the greenhouse relative to the home, and your home's construction. Newer, energy-efficient homes have more issues with humidity because they have fewer air exchanges compared to older, "leakier" homes. Humid air gets trapped inside rather than regularly flushed out. The ideal relative humidity level for homes is in the range of 30%–50%. Here are some tips to keep it in this range, and mitigate damage to the home:

- Add a vapor barrier in the shared wall if there is not already one. This should be placed on the greenhouse side of the wall since that is where the moisture comes from. Additionally, make sure your home siding is in good shape. Normally, this exterior siding of the home is the interior siding of the greenhouse. Older wood sidings in poor condition should not be used in a greenhouse environment; they should be replaced or, at the very least, well protected.
- Protect all surfaces regularly in contact with the greenhouse air (e.g., window and door frames). These should be regularly stained or painted. Or, you can trim them with a more durable non-wood siding material like engineered wood products. See Chapter 9 for siding recommendations.
- Add a humidistat to any fans bringing in air to the home. This will

turn a fan off when the home reaches a maximum threshold for relative humidity.

- Consider adding a heat recovery ventilator (HRV) to the home to create additional air exchanges without overcooling the home. We strongly recommend this if you have a very efficient home with less than ½ air exchanges per hour. See the section in Chapter 7 titled "Season-specific Strategies" for more on HRVs.

Takeaways

- Attached greenhouses present many benefits, such as home heat gain and climate control for the greenhouse.
- Airflow between the home and greenhouse should be controlled to prevent the greenhouse from overheating or over-humidifying the home. Sidings and adjacent surfaces should be well-protected to avoid damage from moisture.
- Controlling air flow depends on where the greenhouse is sited. Thermostatically operated fans offer the greatest level of climate control for the greenhouse and home.

Further Reading

Chiras, Daniel D. *The Solar Home: Passive Heating and Cooling*. Chelsea Gren, 2002.

Shapiro, Andrew. *The Homeowner's Complete Handbook for Add-On Solar Greenhouses & Sunspaces*. Rodale Press, 1985.

Yanda, Bill and Rick Fisher. *The Food and Heat Producing Solar Greenhouse*. John Muir Publications, 1980.

Case Study: Amory Lovins Greenhouse
Greenhouse at the Heart of the Home

900 sq. ft. atrium greenhouse
Snowmass, Colorado

Amory Lovins's "banana farm" is not your typical home. The structure is beyond-net-zero energy (i.e., produces more energy than it consumes), heated mostly by a greenhouse; and grows banana trees high in the Rocky Mountains. The chief scientist and co-founder of the Rocky Mountain Institute renovated the 3,000 sq. ft. home to accomplish this. The original walls were made out of 12"-thick stone. Along with many other materials, they contribute to the large thermal mass of the building. There is so much mass (and insulation) that Lovins predicts

FIGURE 10.3. Amory Lovins Home and Greenhouse. Credit: Tila Zimmerman

if there were a solar apocalypse, cutting off all heat and energy to the home, it would only cool down 1 degree per day.

The greenhouse is a central atrium that serves as the "furnace" for the home. "This 900-square-foot space, plus the heat gain from the other windows, lights, appliances, and people, provides all the heat that's needed for the entire building most of the year. The heat is stored in the masonry, the floor, the water, and the earth under the house. Because of the building's huge thermal capacity, heat is stored for months, not just hours."

With such a large greenhouse space in a tightly built home, excess humidity would normally be a major issue. To handle humidity, Lovins installed a large HRV in the roof. The air-to-air heat exchanger provides efficient ventilation and a source of water. As the humid air cools in the HRV, condensate forms, which is collected for garden watering. In addition, the rainwater from most of the roof surfaces is collected and diverted into large barrels in the greenhouse space.

Another feature that prevents the greenhouse from overheating/overcooling the home is the super-insulated windows; Lovins refers to these as "super-windows." Technically called "krypton-filled Heat Mirror® windows", they work by reflecting radiation back into the space, similar to how low-e films work. According to Lovins, the windows lose only one tenth as much heat as a single pane of glass, and still provide enough light for the established banana trees that give the home its name. "In the first year and a half [the original banana tree] grew to seven meters and gave five bunches of fruit."

Lovins's large integrated structure demonstrates a greenhouse's potential for home heating, but doing it right requires a thorough understanding of the thermodynamics of a building, as well as a willingness to invest in solutions like insulation, higher thermal mass and advanced glazing.

The Lovins greenhouse integrates seamlessly with the home, designed in tandem with the whole structure. If retrofitting a greenhouse onto the home, Mr. Lovins recommends homeowners use a partitioning wall with a sliding glass door to protect the home from excess heat and humidity.

You can find out more about Lovins's home, including all its sustainable gadgets and systems (such as a concrete cantilevered bridge arch for thermal mass) at rmi.org/rmi/Greenhouse.

CHAPTER 11

Earth-sheltered Greenhouses

So far we've discussed how to design and build freestanding and attached greenhouses. Another category deserves attention: underground structures. Underground greenhouses have become increasingly popular in recent years, driven partly by interest in the "walipini," a style of pit greenhouse that has sparked plenty of social media buzz. While online pictures and headlines make it look easy, building a greenhouse underground takes careful planning and construction.

Why Build Underground?

Underground greenhouses are naturally more energy efficient than above-ground structures but the reason is sometimes confusing. Soil is actually a very poor insulating material by itself. Soil provides some insulation, but is primarily effective because it experiences less temperature fluctuation, particularly as depth increases. This reduces the temperature differential between inside and outside, called delta T. As delta T decreases, so does heat loss.

It is important to realize, though, that this does not make the surrounding soil warm. Your target greenhouse air temperature will probably be higher than the surrounding topsoil. Thus, even underground greenhouses require insulated walls. Soil helps insulate some, but is not a substitute.

Hot climates, like areas of the Southwest, can be an exception. In these areas, the soil may be closer to your ideal indoor temperature, in which case the soil helps keep the structure cool. Use insulation underground if there is a significant difference between your ideal indoor

Case Study: Garden Pool
The Swimming Pool Greenhouse

480 sq. ft. residential greenhouse
Mesa, Arizona

When Dennis McLung and his family moved into a foreclosed home in Mesa, Arizona, it came with an empty, dilapidated swimming pool. Instead of a water-intensive luxury, they saw it as a potential year-round garden. Using a wood frame, steel cable rafters and UV-stabilized polyethylene film, the pool became a low-cost underground greenhouse. The deep end is 6' underground, tapping into constant earth temperatures that allow the greenhouse to grow year-round through the scorching desert heat. The greenhouse stays 10–15 degrees cooler than outside, while using a fraction of the water an outdoor garden would need. The McLungs integrated aquaponics and chickens into the greenhouse, creating a series of intertwining resource loops they call the Garden Pool (more information at gardenpool.org).

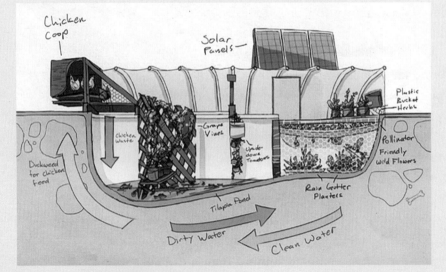

FIGURE 11.1. Credit: Garden Pool

conditions and the soil temperatures (which you can look up using the National Resource Conservation Service's [NRCS] Soil Climate Analysis Network [SCAN] resource). See the section in Chapter 6 titled, "Insulating Underground" for more on methods of underground insulation.

Challenges of Building Underground

Building any structure underground comes with significant challenges:

- **Water and drainage:** Drainage is a major issue with underground structures. Water can either slowly seep from the sides or the floor, or it can drain off the surrounding land during rains. The water table—the depth at which the ground is fully saturated—is another factor. The floor of the greenhouse should be a few feet above the water table, which may make underground structures prohibitive in some locations.
- **Durability:** Underground greenhouses *can* be durable, but it takes some extra steps to make them so. As the saying goes, nature abhors a vacuum. When creating an empty hole in the earth, the walls of the structure need to be re-enforced to prevent them from caving in. They will be under greater pressure than a freestanding greenhouse's walls, particularly if your soil freezes in the winter.
- **Site prep and building costs:** Underground greenhouses require much more work for site preparation—excavation and building retaining walls. Underground greenhouses *can* be cheap if you provide the labor and use recycled materials; more often, though, underground structures take longer to build and cost more than aboveground greenhouses due to the added site prep.
- **Shading:** Often overlooked by cursory articles or blogs, shading is very important to consider when digging into the earth. Underground greenhouses often have low roof angles due to their geometry, and there can be little above-ground glazing. This can shade some or all of the growing area, and creates problems with snow shedding. As we'll detail below, it is critical to think about solar angles and the geometry of the structure to ensure full light.

"Walipinis"

There are a few different styles of underground greenhouses. The most basic is a *pit greenhouse*: a hole dug in the ground with glazing laid over it. Recently, these have been called "walipinis." The name stems from a project organized by the Benson Institute (a research arm of the Mormon Church). In 2002, volunteers designed and built several pit greenhouses in the mountains of Bolivia. They called the design a walipini, which means "place of warmth" in the native language. A walipini is a simple 6'–8' hole dug in the ground covered with a double layer of polyethylene glazing. The builders' intention was to create a low-cost, year-round growing structure for local farmers.

Several years later, the idea of a walipini has picked up a surprising amount of buzz. A popular *Tree Hugger* article proclaims that you can "build an underground greenhouse for $300" based on the walipini design. Unfortunately, the article overlooks how many aspects of the design need to be modified in order to work in North America. Namely, if you consider the light angles for most latitudes in North America, it becomes apparent that most of the floor area of a pit greenhouse will be shaded for most of the year by the front wall. The roof angle works well in Bolivia (16 degrees latitude), where the angle of the sun is much higher through the growing season, but would shade most of the plant area at higher latitudes.

For this reason, we don't recommend the basic pit greenhouse design as a year-round greenhouse in North America. Beyond shading issues, the lack of reinforced walls is a major problem for durability. In climates with soaking rains, freezing winters, and high snow loads, digging a simple pit in the earth is a great way to get a mud swimming pool, not a greenhouse.

Earth-sheltered Greenhouses

To accommodate for the low sun angles farther from the equator, an "earth-sheltered" greenhouse is more appropriate for North American growers. In this design, the north, east and west sides are buried partially underground. The south side includes some glazing and is often partially above ground.

Building a Quality Underground Greenhouse

The greenhouse at Raven Crest Botanicals in upstate New York demonstrates the quality building necessary to create a durable underground structure. Farm owner Susanna Raeven and builder Jesse Patrick excavated 2' underground for the top layer of the structure, and 4' underground for the cold-sink. It's constructed with posts set in concrete and sided with white oak. The bottom half, in contact with the soil, is protected with pond liner and a breathable under-layer called Cedar Breather. After construction, they created berms around the north, east and west of the greenhouse using soil and crushed gravel. Today, you can just see the roof and south-facing glazing poking out above ground. Below ground is a lush, sun-filled space that grows and dries medicinal herbs for the organic farm. Susanna has written a blog for *Mother Earth News* about the project, which includes many more photos and details. See "How to Build an Earth-Sheltered Greenhouse" at goo.gl/4mhfRv.

FIGURE 11.2.
An Earth-sheltered Greenhouse.
Credit: Raven Crest Botanicals

The "Dirt-cheap" Underground Greenhouse

Many of these designs utilize south-facing hillsides, creating a natural berm around the north side of the greenhouse. However, they can also be built into shallow-sloped or flat sites by making the greenhouse wider, and having some of the front area intentionally shaded. This strategy was popularized by Mike Oehler, author of *The Earth-Sheltered Solar Greenhouse Book*, among other books.

Mike uses the front portion of the greenhouse as a "cold-sink," an area buried deeper than the rest of the growing area, as shown on the cover of his book, Fig. 11.3. The cold-sink does two things: first, it collects cold air, allowing warmer air to rise to plant level during the day. (Mike keeps rabbits here, which tolerate the cooler temperatures and serve as a small heating element). Secondly, the cold-sink creates a geometry such that the growing area is illuminated year-round. Some of the south wall typically extends above ground to create the necessary glazing angles.

When designing an earth-sheltered greenhouse, it's important to sketch out the solar angles at different times of year and predict how they will interact with the greenhouse to avoid shading. This way, an earth-sheltered greenhouse can have the added efficiency of an underground structure, as well as sufficient light for year-round growing.

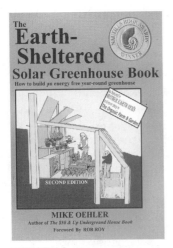

FIGURE 11.3.

Earth-sheltered Greenhouses

Mike Oehler bought his property in northern Idaho in the 1970s, where he lived and grew his own food in underground structures for over 30 years. (He passed away in February, 2016, leaving behind a legacy of building low-cost underground structures which he documented in his several books). An incredibly frugal person, Mike built his earth-sheltered greenhouse almost entirely from scrap materials. The walls are made out of locally harvested wood from his land. They are wrapped in a layer of polyethylene plastic to protect the wood from moisture. Using his own materials and labor, Mike proudly noted that the greenhouse cost him only $380 to build.

Construction Methods

Before construction, the greenhouse site must be excavated with a backhoe or excavator. If you have a flat site and are using the cold-sink concept, the front area is dug 2'–4' deeper than the top floor of the greenhouse, depending on the slope of the site and dimensions of the greenhouse.

Building underground is more complicated due to two major challenges: added pressure from the surrounding soil, and moisture. Unprotected wood will rot if left in contact with the soil. Thus, underground greenhouses require extra measures to deal with moisture and added loads from soil. There are several construction methods to choose from.

Post and Beam Construction

Post and beam construction is a popular method for underground greenhouses, mostly because it's the least costly. As described in Chapter 9, posts (usually 6" × 6") are set deep into the ground and backfilled with concrete. (Or the posts are bolted to concrete piers that are poured underground.)

The challenge with post and beams is protecting the wood from moisture in the soil. There are a few options to do this. "Post protectors" are slide-on pieces of plastic (either PVC or HDPE) which cover a 6" × 6" post. The other method to wrap posts in a waterproof material like EPDM. EPDM, commonly called pond liner, is a thick plastic material. It costs less than rigid plastic protectors, but it can tear. In either case, there is still the risk of water getting between the barrier and the wood—an inherent risk of underground structures.

After the posts are set, usually a few feet apart depending on the size of the structure and pressure from the soil, walls are constructed out of wood or water-resistant siding. Again, wood used underground will need to be protected with a waterproof barrier like pond liner on the exterior, shown in Fig. 11.4. Between the pond liner and the wood should be a layer that carries water away from the wood, and allows the wood to dry out if it gets wet. These materials are called a *breathable underlayment*, often used under wood shingle roofing. We have used a

FIGURE 11.4.
Earth-sheltered
Greenhouse.
Credit: Raven Crest
Botanicals

slightly different material called a *vertical wick drain*, which drains water away from the structure.

After the posts are set, the roof rafters are aligned with the posts. The rafters brace the posts, preventing them from caving in. Since these need to be perfectly aligned, it is important to lay out the whole construction plan early on. See *The Earth-Sheltered Solar Greenhouse Book* for more on post and beam construction. We also recommend talking to a structural engineer about your design.

Concrete Retaining Walls

The most durable method of building an underground structure is to build concrete retaining walls. Concrete will hold up to downward pressure of the soil and create a clean, finished room inside.

This also allows for some flexibility in building the above-ground structure. On top of the concrete wall, you can use stick framing or even structurally insulated panels, as shown in Fig. 11.5. The downside with concrete is that it is an expensive construction process. Typically, the retaining wall includes support walls on either side. After these are set, a contractor must come back and pour the retaining wall. That work spans a few days and comes with a considerable price tag. A structural engineer should specify the wall layout and thickness.

FIGURE 11.5.
An earth-sheltered greenhouse is constructed out of a concrete retaining wall and Structurally Insulated Panels (SIPs). A photo of the finished greenhouse appears in the color section.
Credit: Ceres Greenhouse Solutions

Concrete Blocks or Insulated Concrete Forms

Another method is stacking standard concrete blocks, reinforcing them with rebar, and filling them with poured concrete. Alternatively, blocks can be insulated concrete forms, discussed more in Chapter 9. Like a concrete retaining wall, these provide a long-lasting durable wall, as well as a nice finished aesthetic on the inside. They also give the greenhouse added thermal mass inside the structure.

Moisture and Drainage

There are a few different methods of dealing with moisture in an underground structure. For more on these, we recommend researching moisture management in earth-sheltered homes or basements (see Further Reading). Essential tactics include:

- Use waterproof barriers around walls to prevent water from seeping in from the surrounding earth.
- These can be as simple as a gravel-filled hole in the center of the greenhouse. As water drains down the pit, it gradually seeps further into the soil underground. A more advanced method is installing a

sump pump. This involves burying a sump basin underground for water to drain into.
- Install floor drains.
- Use French drains installed around the edges of the greenhouse. These drain water away from the structure. French drains are simple trenches filled with gravel, with a perforated drain pipe in the center. This reduces water runoff into the structure during rains.
- Finally, it's helpful to landscape sloped berms around the greenhouse to help drain water away from the structure. You can also backfill around the walls with a well-draining soil or gravel.

When it comes to drainage, and underground building in general, keep in mind that the factors involved are strongly influenced by your local soils. This is an area where it is wise to consult a local contractor and structural engineer if you do not have experience.

Takeaways

- In order to function as a year-round growing environment, most underground greenhouses in North America (or far from the equator) should have an earth-sheltered design, with walls partially underground. Unless the greenhouse is built into a hillside, the glazing usually needs to extend above grade to get full light across the greenhouse floor.
- South-facing hillsides are excellent sites for earth-sheltered greenhouses.
- Many additional factors need to be considered in the construction process, such as reinforcing the walls, drainage, and shading.

Further Reading

Oehler, Mike. *The Earth-Sheltered Solar Greenhouse Book*. Mole Publishing, 2007.
Benson Agriculture and Food Institute. *Walipini Construction (The Underground Greenhouse)*. Brigham Young University, Revised 2002.
Many of the same principles that go into building an earth-sheltered home apply to greenhouses as well. We recommend resources on earth-sheltered construction such as Rob Roy's *Earth-Sheltered Houses*. New Society Publishers, 2006.

HEATING AND COOLING METHODS

CHAPTER 12

Passive Thermal Mass

On most sunny days, even during the winter, there is a surfeit of solar energy entering the greenhouse. In other words, all the Btus of energy received during the day are enough to keep the structure warm over a 24-hour period. The challenge is timing: heat comes during the day, often excessively, but is mostly required at night. Many of the following heating/cooling systems function by simply storing the excess heat during the day for use as heating at night. Though they facilitate heating and cooling, really they are simply *thermal storage* strategies. To that effect, thermal mass materials are the oldest and simplest strategy for storing heat, and naturally mitigating temperature swings.

How It Works

When light hits a material, some of it is converted to heat. Thermal mass materials absorb this heat via conduction. Heat slowly conducts from the surface to the center of the mass, allowing the entire volume to heat up by a few degrees in a day. When the air temperature in the greenhouse drops at night, the mass slowly starts radiating this heat through conduction and by releasing short-wave infrared radiation (thermal radiation). In this way, thermal mass regulates the temperature of the greenhouse. It absorbs excess heat during the day and slowly re-radiates

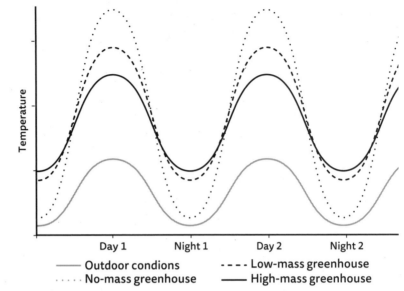

FIGURE 12.1.
Effect of Thermal
Mass on Temperature.

it at night (or whenever the air temperature drops below the temperature of the mass), evening out the daily temperature swings.

Because mass materials store heat, they are often referred to as "heat sinks," "heat banking" or "thermal batteries." The idea is the same: store as much heat as possible when it is over-supplied (the day), and slowly distribute it when it is in demand (at night or on cold and cloudy days). To do this, it's important to maximize its exposure to light in the winter (or whenever greenhouse heating is needed). Like a battery, this allows it to charge as much as possible during the day, so it can radiate this heat later. Even if the mass is not directly illuminated, it still absorbs some heat from surrounding air. This allows it to have some effect on overcast days; however, the predominant driver of a mass's ability to store heat is direct light.

During the summer, the goal of mass is to help keep the greenhouse cool. To that end, mass should be shaded during the summer months. On summer days, it will still absorb some heat from the greenhouse as hot air moves across it. It then radiates this heat at night, given a sufficient drop in temperature, and resumes cooling the next day.

Choosing a Thermal Mass Material

All materials absorb and store heat to some extent, but some are much more effective than others. The amount of heat a material can store is called its *heat capacity*, and is determined by two factors. First, the material's *specific heat* is a property of the material, defined as how much energy is required to raise its temperature 1 degree. Metric units are joules/gram/Celsius. The imperial units, standard in the US, are Btu/lb./Fahrenheit.[1] Both describe how much energy (joules or Btu) must be applied to a unit of mass (grams or lbs.) to raise it one degree in temperature. As slow-adapting North Americans, we'll continue to use imperial units going forward.

Density is a second factor that affects a material's heat capacity. The more mass in a given volume, the more energy it can store. Most mass materials, like stone and concrete, are effective because they are dense.

When you multiply these two characteristics—specific heat and density—the result is a material's *volumetric heat capacity*. We find this is the most useful unit of measurement because it indicates the total heat stored in a given volume of the material. In a greenhouse setting, volume is an extremely important factor, since mass materials can take up considerable space otherwise used for growing. Figure 12.2 gives the volumetric heat capacities of common thermal mass materials. Air is also given for comparison.

Note that heat capacity is a different property than *conductivity*, the rate at which a material conducts or transfers heat. Something that often confuses people is the fact that thermal mass materials are not good insulators, generally speaking. When a material is a good insulator it resists heat flow. (The "R" in "R-value" stands for resistance to heat flow.) Thermal mass materials can *store* a lot of heat, but easily conduct heat with their surroundings. This means they should be insulated from the exterior and should not substitute for insulation.

The units in Fig. 12.2 are Btu per ft^3 per degree Fahrenheit. This number represents how much energy the material *can* store. How much energy it *does* store depends on its exposure to light, heat and other

FIGURE 12.2. Volumetric Heat Capacities of Common Building Materials.

Material	Heat Capacity per Volume (Btu/ft³/F)
Water*	62
Steel	59
Rock or Stone (30% air gaps)	25
Soil (damp light soil)	25
Concrete	25–32
Brick	23–25
Adobe	20
Wood (pine or fir)	18
Sand	18
Air	0.018

*Helpful conversion: 1 gallon of water has a volumetric heat capacity of 8.34 Btu.

factors. A water barrel sitting in the sun all day will store much more energy than one in the shade. Figure 12.2 is useful for comparing different materials. Calculating the total heat storage effect is a more complicated equation we'll return to in "Sizing Passive Mass," below.

Figure 12.3 shows the common ways builders incorporate these materials into the greenhouse. Every greenhouse inherently contains some thermal mass in its structure, soil, water and plants. While not insignificant, these sources of mass are usually not enough to provide year-round climate control. Thus, thermal mass materials are integrated in larger quantities, usually via water or masonry.

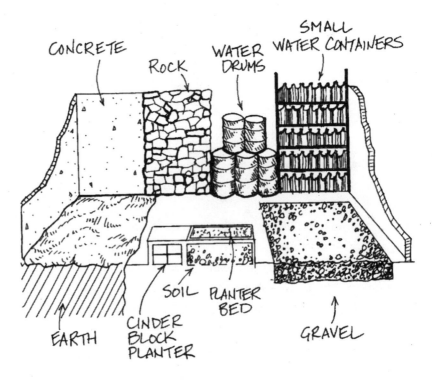

FIGURE 12.3. Common Thermal Mass Materials in a Greenhouse.

Water as Thermal Mass

By far and away, water is the most commonly used thermal mass in greenhouses—for clear reasons. It has the highest heat capacity per volume of any of readily available material, and it's cheap. Stacking several large drums of water on the north wall of the greenhouse creates a "water wall" a large, low-cost thermal battery.

Water walls should be exposed to light in the winter and shaded in the summer. If water barrels are on the north wall of the greenhouse, this is easily done with a partially insulated roof, as shown in Fig. 12.4. The length of the insulation determines when the water wall is exposed to light (heating season) or shaded (cooling season) based on the solar angles at your location. To find the right length of insulation, sketch the greenhouse profile, the dimensions of the water barrels, and the solar altitude angles during the summer and winter solstices (and/or equinoxes). By using a program like SketchUp, or a protractor, you can estimate the length of the roof insulation that allows the mass to be shaded in the summer, and fully illuminated in the winter.

Though it takes a little longer to build, we strongly recommend adding this section of roof insulation. Not only does it enhance the effect of the thermal mass, but it also reduces heat loss through the roof (where heat loss is greatest).

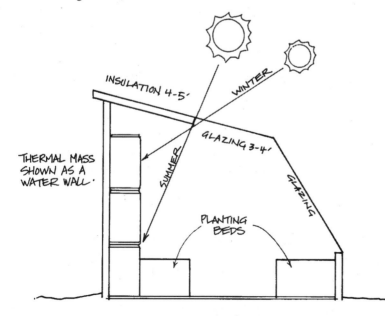

FIGURE 12.4.
Sun Angles
and Thermal Mass.

Is it better to have large or small containers?

Large containers allow you to efficiently incorporate more water into the greenhouse space. They also have a lower surface area relative to their volume compared to small containers. Surface area determines how quickly a material can conduct heat with the surrounding air. To illustrate, let's compare one 55-gallon barrel to 55 individual gallon milk jugs. While both contain 55 gallons of water, the large drum has a smaller surface area relative to its volume. Because it has a smaller area to conduct heat, the large drum takes much longer to heat up and cool down than the individual jugs. It has a smaller effect on the greenhouse temperature in the short-term, but is able to store heat over a longer period. Thus, larger containers are more useful to keep the greenhouse above freezing over prolonged cold and cloudy periods. Milk jugs of water will create a quick boost in temperature, but will likely freeze after prolonged freezing temperatures. Thus, if water is the primary heat storage mechanism, most growers go full-bore and use large containers. Moreover, they are also easier to stack and store, allowing large volumes of water to be placed space-efficiently. Small containers can also be used for more localized protection around sensitive plants; they have the benefit of being more flexible with placement.

Selecting Water Containers

The most common containers are 55-gallon barrels, either plastic or steel. Steel barrels are better for performance since metal is more conductive than plastic (i.e., they are able to transfer heat more quickly). However, they are also prone to rusting, so they have shorter lifetimes. Barrels usually rust from the outside first, as water collects in the lid of the barrel and continually evaporates off the surface. If using steel, keep barrels off of wet soil and consider using plastic coverings on top of the barrels.

Both plastic and steel barrels can be found cheaply and reused. Landscaping companies, your local roads/highway department, scrap yards, container companies and Craigslist are good sources for low-cost barrels.

Water walls are simple in concept, but come with a serious need for caution. A full 55-gallon water barrel weights almost 500 lbs.... heavy

FIGURE 12.5.
Water Wall at the
Cheyenne Botanic
Gardens.

enough to crush you if it were to fall. Water walls need to be stable and structurally supported. They should be on a level surface, either concrete or flagstone. Longtime solar greenhouse designers Penn and Cord Parmenter advise building in a 2' wide concrete slab under the barrels, while the rest of the floor can be open to the soil for planting. Additionally, if stacking barrels, use stable shelving in between layers, like the painted plywood shelves shown in Fig. 12.5. Penn and Cord go one step further and use wood straps across barrels to ensure they don't topple.

Other Tips for Using Water Barrels

- **Leave an air gap:** After barrels are installed in the greenhouse, fill them on-site. Leave an air gap since water will expand slightly as it warms up during the day.
- **Avoid freezes:** A large water container will likely break if it freezes and the water expands. Freezing a 55-gallon drum of water is rare, because water gives off a great deal of heat as it changes from a liquid to a solid (an example of the energy transfer of phase changes discussed later in this chapter). However, it can happen if the greenhouse is left open through the winter—something to consider if you are a three-season grower. Smaller containers are more prone to freezing.

- **Paint drums a dark color:** Darker colors absorb more heat. That means they reflect less light, but maximizing heat absorption is a primary goal of thermal mass. Water containers should be painted a dark color, unlike the rest of the greenhouse interior, which is painted white to reflect light. Conventional wisdom says to paint drums black. Interestingly, studies show that a dark blue or red color absorbs almost the same amount of heat, and reflects light in the red or blue spectrum, which are more useful for growth.
- **Plan thermal mass into your floor plan:** A standard 55-gallon drum is 2' in diameter. It's critical to plan the location and spacing of mass when designing your greenhouse floor plan so you don't end up with too little growing room.

Masonry as Thermal Mass

The second most common method of incorporating passive thermal mass is to add masonry materials, such as concrete, stone or brick. These have the advantage that they can serve other purposes, such as a foundation, retaining wall, raised beds, flagstone pathways, etc. The downsides are that they have a much lower heat capacity per volume, and they are much more expensive. We recommend using them strategically, taking advantage of their structural or aesthetic benefits, but not solely for the purpose of thermal mass. For example, some growers opt to put in a concrete slab floor/foundation if they will be growing with hydroponics, aquaponics or rolling benches. The concrete contributes significantly to the overall thermal inertia that resists temperature swings. (It often called a "solar slab" when used in passive solar homes). However, due to the high cost, a concrete slab is not a wise choice if used solely for the effect of thermal mass.

If you do choose to incorporate masonry, you can pick up some pointers from passive solar homes (see Further Reading). The Sustainable Business Industry Council recommends thermal mass walls be no more than 4"–6" thick, depending on the material, to allow for adequate heat conduction. This returns to the same principle that dictates the use of large or small water containers: the surface area relative to the volume should be great enough that the mass can conduct heat to the air. As you increase the volume to surface area ratio (with thicker walls), there are diminishing returns.

Sizing Passive Mass

How much water or concrete does the average greenhouse need to stabilize temperature swings? A wide array of design variables makes this different for every greenhouse. However, over the years a general rule of thumb has emerged: use 2–5 gallons of water per square foot of glazing area if thermal mass is the primary climate control strategy. Colder climates should be on the higher end of this range and warmer climates on the lower end. The recommendation is relative to glazing area, as this is a better indicator of the greenhouse's heat gain and heat loss, rather than the footprint. To evaluate other materials, simply convert this rating—given the volumetric heat capacity of water—to another material using the figures in Fig. 12.2. Concrete, for instance, has about half the heat capacity of a volume of water, so twice as much is needed to achieve the same effect.

As you can tell, the above is a very general recommendation; it has to be, because there are so many factors that go in to sizing thermal mass: the climate, the design, the mass material used, the container it is in, and where it is used. The rule of thumb is intended to keep greenhouses above freezing, and indeed, it has worked for greenhouses in a range of climates as shown in several examples in this book. Most of the examples that rely on thermal mass are in cold and sunny climates like Colorado and Wyoming; however, mass has been successfully used in cloudy climates as well. The New Alchemy (now Green Center) greenhouse in Cape Cod, Massachusetts, for example, integrates mass in every way possible: it holds close to 10,000 gallons of water in several large fish tanks (shown in the color section), as well as large rock retaining walls, flagstone paths, cement foundation walls, soil, etc. Combined, all the thermal mass is able to keep the 1,800 sq. ft. solar greenhouse above freezing without backup heat through the cold and cloudy winters. Their typical indoor lows are about 40°F (4°C), despite outdoor lows of −10°F (−23°C).

A more specific determination of how mass affects the indoor environment requires an in-depth energy analysis using energy modeling software. We conducted one such analysis for a hypothetical greenhouse located in Boulder, Colorado, varying the quantity of water (in gallons) in the greenhouse while keeping all other variables constant. Figure 12.6 shows gallons per square foot of glazing area on the horizontal axis. The vertical axis shows the number of hours over the course of a year in which the greenhouse will experience temperatures greater than 90°F (32°C), between 50°F (10°C) and 90°F, or less than 50°F, depending upon the number of gallons used. You can see that by adding more water, the number of hours the greenhouse is over-heated (>90°F) or under-heated (<50°F) goes down, and the number of hours it stays within an ideal growing range (50°F–90°F) goes up. Figure 12.6 shows that, like many variables in the greenhouse design, there is a great benefit with adding some mass, but diminishing returns as you add more and more.

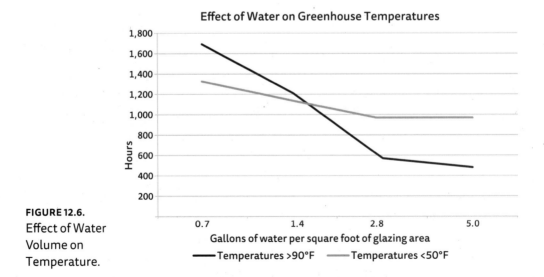

FIGURE 12.6.
Effect of Water
Volume on
Temperature.

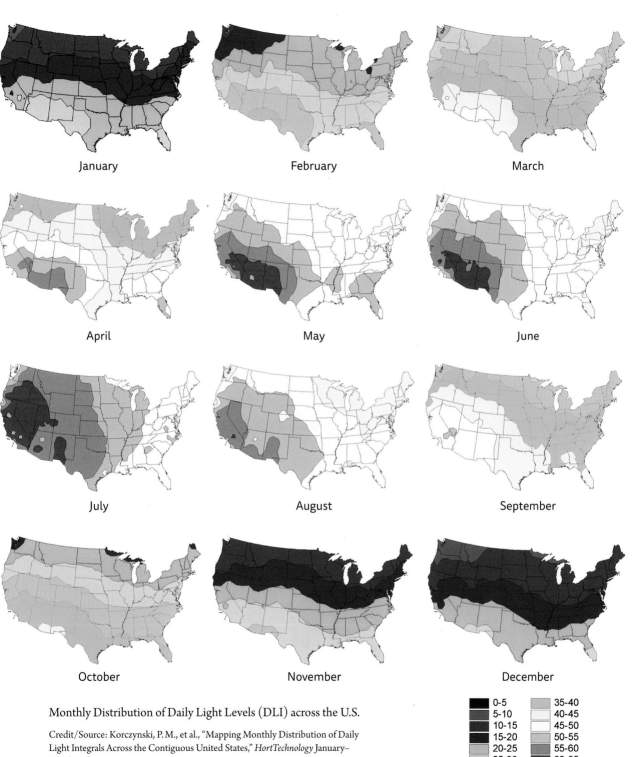

January

February

March

April

May

June

July

August

September

October

November

December

Monthly Distribution of Daily Light Levels (DLI) across the U.S.

Credit/Source: Korczynski, P. M., et al., "Mapping Monthly Distribution of Daily Light Integrals Across the Contiguous United States," *HortTechnology* January–March, vol. 12:1, 12–16, 2002.

0-5	35-40
5-10	40-45
10-15	45-50
15-20	50-55
20-25	55-60
25-30	60-65
30-35	65-70

Above: Verge Permaculture

Above right: A-Frame Greenhouse designed by Whole Trees Architecture

Right: Underground Greenhouse by Raven Crest Botanicals

Below: The Golden Hoof Farm

Above left: Dirt Craft Natural Building

Above: Amory Lovins's Atrium.
Photo by Tila Zimmerman

Left: Earth-sheltered Greenhouse.
Raven Crest Botanicals

Below: Hillside Greenhouse by
Ceres Greenhouse Solutions

Above, below: Cheyenne Botanic Gardens

Above right, right: The Green Center

Above left: Jasper Hill Farm

Above: Photo by Craig Schiller

Left: Greenhouse and Chicken Coop
by Ceres Greenhouse Solutions

Below: Aquaponics Greenhouse at The Sage School

Above: Growing Power

Above right: Aquaponics Greenhouse by Ceres Greenhouse Solutions

Right: Climate Battery at The GrowHaus

Below: School Greenhouse designed by Ceres Greenhouse Solutions

Above left: Penn and Cord Parmenter/
Smart Greenhouses LLC

Above: Ceres Greenhouse Solutions

Below: Greenhouse by Jack Coddington

Above: Photo by Jack Coddington

Above right: Flourish Farms

Right: Rocket Mass Heater at
The Golden Hoof Farm

Below right: Concrete pier foundation
with perimeter insulation

Below: Attached greenhouse with
rain water catchment

Pros, Cons and Best Applications

The primary advantage of conventional thermal mass materials is that they are low cost (in the case of water) and can serve other useful functions (in the case of masonry). They are also simple to install and maintenance-free. Most importantly, they don't require any electricity, helping facilitate a year-round passive solar greenhouse.

The main drawback is that they are bulky. A water wall takes a large amount of potential growing room in a residential greenhouse. Though cheap in materials, integrating a water wall can require building a slightly larger greenhouse.

Secondly, thermal mass materials offer limited control because they are strongly tied to daily weather fluctuations. Passive thermal mass relies on sufficient direct light during the day to have a pronounced effect. This limits its effectiveness in cold and cloudy areas, like parts of Canada or the northeastern US.

Passive mass is also less useful in hot climates that do not experience significant daily temperature fluctuations like the southern US. If the greenhouse does not cool down enough at night, the mass will simply stay warm, and won't be able to absorb heat the next day.

A final drawback is that mass materials do not distribute heat very evenly throughout the greenhouse, but instead create a pocket of warm air directly around the mass. Many growers report crops near the mass benefit greatly, while those farther away are still susceptible to freezing or overheating. Growers can modify their planting plan to suit these growing zones, or use fans to help distribute the heat throughout the greenhouse.

Phase Change Materials (Thermal Mass 2.0)

As noted above, a primary downfall with incorporating large volumes of water is simply sacrificing growing area. What if you could create the same thermal storage effect using only a fraction of the space? *Phase change material* (PCM) makes that possible by harnessing the chemical reaction of phase changes from liquids to a solids. PCM passively stores heat, but has a much greater heat capacity per volume. For that reason, we often call it "water 2.0."

How It Works

To understand how PCM works, we first must look at the energy transfer caused by phase changes.

Fact: It takes 1 Btu to heat 1 pound of water 1 degree from 30°F to 31°F (−1°C to −.6°C), but 80 Btu to heat water from 32°F to 33°F (0°C to .6°C). Though both only rise 1 degree in temperature, the latter requires 80 times more energy to do so. The reason is that when you heat water from 32°F to 33°F, it melts. As molecules change from solid to liquid, it requires much more energy, which the water absorbs from the surrounding environment. When the phase change occurs in the opposite direction, freezing liquid water back into solid ice, energy is released.

Another way to understand this process is to see the effect on water's temperature as you apply heat, as shown in Fig. 12.7. Let's say there is a pot of water sitting on the stove, which is applying heat at a constant rate. The temperature steadily rises until the water reaches the boiling point. When water boils and starts to evaporate, the temperature flatlines. The added heat is absorbed by the phase change (called an *endothermic* reaction). When the water vapor condenses back into a liquid, the opposite occurs: it releases heat (called an *exothermic* reaction). In effect, the phase change "stores" the energy through the latent heat of evaporation.

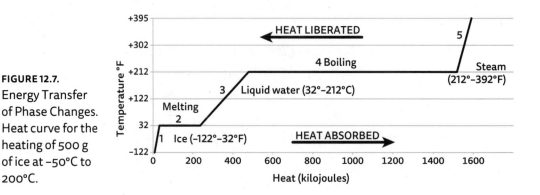

FIGURE 12.7.
Energy Transfer of Phase Changes. Heat curve for the heating of 500 g of ice at −50°C to 200°C.

Phase change materials take advantage of this basic process for a specific purpose by melting/freezing at a specific temperature. The most basic PCM is water, used to regulate temperatures around water's freezing point. For example, a longtime farming practice is spraying crops with water on freezing cold nights. As the water freezes, it gives off heat, helping keep the temperature around the plants from dropping further below freezing.

The problem is that water changes phases at inconvenient temperatures (32°F/0°C and 212°F/100°C). Phase change materials are engineered fluids that are designed to "freeze" (or solidify) at more moderate temperatures. For greenhouse applications, this is typically 50°F–70°F (10°C–21°C)—the temperature range ideal for a growing environment. Phase change materials serve the same purpose as passive thermal mass—they absorb heat and later radiate it, helping smooth temperature swings. However, they work quite differently, by using the energy transfer of a chemical phase changes. Passive mass, in contrast, stays in one state. In this way, phase change materials are able to absorb, and radiate much more energy compared to the same volume of passive mass, thus providing the thermal mass effect without taking up much space.

Let's pull this all together with an example. Say you purchase 200 sq. ft. of the PCM with a capacity of 91 Btu per square foot. (As explained below, a common PCM for greenhouses is a rated in square feet, not volume). You install this on the north wall of your greenhouse, yielding a total capacity of 18,200 Btus. You select a material that is designed to change phases at 65°F (18°C), also called its *set temperature*.

When the greenhouse heats up during the day and crosses the 65°F threshold, the PCM melts and absorbs 18,200 Btus from the air, helping cool the greenhouse. At night, when the greenhouse drops below 65°F, it radiates 18,200 Btus back into the air, heating the greenhouse.

FIGURE 12.8.
BioPCM Mat. Credit: Phase Change Energy Solutions

The PCM in this example can be housed inside a wall cavity, so it doesn't take up growing room in the greenhouse. In comparison, to achieve the same storage effect with water barrels would require several 55-gallon drums of water.[2] Five barrels of water would occupy about 10 sq. ft. of the greenhouse, a significant amount of growing room in the hypothetical 120 sq. ft. greenhouse. This large heat storage capacity combined with space-efficiency are the prime advantages of using PCM in a greenhouse.

History of Phase Change Materials

The energy transfer caused by phase changes has been used for centuries, and is responsible for many everyday technologies. Notably, a refrigerator works by changing the pressure of a liquid and converting it from a gas to a liquid and back again. Evaporative coolers, geothermal heat pumps, and many other systems work by harnessing the energy transfer of phase changes. Using phase changes via materials (i.e., PCMs) has been used for decades, most commonly in the medical industry. More recently, PCMs have become more cost-effective and practical to use in buildings. As an emerging product, PCMs are still rare in greenhouses. However, given the need for thermal storage and space-efficiency, they hold tremendous potential for the industry.

Sourcing and Installing PCM

Traditionally, phase change materials have been made from salt hydrates (salt-water mixtures) or paraffin waxes (petroleum-derived products). Both have high storage capacities (also referred to as a material's *latent heat*) and freeze/melt at a range of temperatures. However, they both have practical challenges: salts are highly corrosive, and paraffin waxes are expensive in large volumes, plus they have a large carbon footprint.

A third category of PCMs are those made out of oils derived from plants. Called "bio" or "organic" PCMs as a category, these are more cost-effective and environmentally sustainable for large applications. The main product on the market for building use, called "BioPCM," made by Phase Change Energy Solutions, is shown in Figs. 12.8 and 12.9. Its main advantage is that, as a thin flexible mat, it can be installed inside the wall cavity. We've also seen clients use it in existing greenhouses (installed on top of the north wall), cold frames, hoop houses, around aquaponics fish tanks…it has a huge range of applications where added temperature regulation is needed.

BioPCM comes as a rollable mat, about ½" thick and 16" or 24" wide, so it is sized to fit a typical stud bay. The material is housed in pods, which maximize the surface area for adequate heat conduction. The edges of the mat are easily stapled onto the studs of a conventional stick frame greenhouse, as shown in Fig. 12.9. BioPCM comes in a range of products that vary by cost, the set temperature, and their heat capacities. They currently cost $5–$15 per sq. ft., depending on the heat storage capacity of the material. For example, you can buy a 50 Btu/sq. ft. or a 200 Btu/sq. ft. product. Higher heat capacities are more expensive, so this decision comes down to your budget and goals. If you only have a very small area in the greenhouse, you probably want to upgrade to a higher heat capacity material.

Many of the same recommendations for passive thermal mass materials apply to PCM. First, it should be in an area directly exposed to light, so it can conduct enough heat to take effect every day. We often install it on the north wall of a greenhouse. Though we say "exposed to direct light," we recommend covering products like BioPCM in a thin, conductive siding material. Currently BioPCM is packaged in a plastic

FIGURE 12.9.
BioPCM Installation.
Credit: Phase Change
Energy Solutions

film that is susceptible to UV damage. If the packaging fails, the PCM loses its effect. (Made out of plant oils, bio-based PCMs are nontoxic and harmless to the garden if they spill.) A thin plastic or metal siding keeps the product out of direct light, but still able to conduct heat to the greenhouse.

Just like water, PCM should be installed on the interior side of the wall with insulation behind it so it can only conduct heat into the greenhouse. It requires a sufficient surface area to conduct heat and take effect every day. Long thin packages like the examples shown here allow the product to absorb enough heat to undergo a phase change at the appropriate temperature.

A critical factor is the set temperature of the PCM product (the temperature at which the material melts/freezes). Manufacturers are able to customize this for the application. So, you have flexibility in choosing a set temperature based on your desired indoor temperatures.

The set temperature should also be based on the predicted temperature range of your greenhouse, since the greenhouse must cross this threshold every day for it to take effect. For example, in cooler climates, we recommend a PCM with a lower set temperature of 50°F–60°F (10°C–16°C). Hot climates will logically use a higher set temperature (e.g., 80°F [27°C]) to ensure that the greenhouse cools enough at night in order for the PCM to reset and take effect the next day.

A final consideration is the material packaging and durability. When it comes to the lifetime of the PCM material, this generally comes down to the packaging, which is often plastic. Salt hydrates will slowly corrode materials, reducing their lifetime. Bio-based products last longer—10 to 20 years—as long as the packaging does not fail due to tears or by getting brittle under UV radiation.

Because many factors go into the effectiveness of PCM—the packaging, product, set temperature and location in the greenhouse—we recommend talking to a supplier or distrib-

Factors When Selecting a Phase Change Material

- Cost per volume
- Latent heat storage (the heat storage capacity, rated in Btus)
- Temperature at which the material reacts (the "set temperature")
- Packaging materials (durability, UV resistance)
- Surface area relative to volume of PCM
- Source of the material and shipping costs

utor about the specifics of your project. Not all materials are intended for greenhouse use, and getting the installation and application right determines the effectiveness of the PCM.

Pros, Cons and Best Applications

The primary advantage with PCM is its ability to store large amounts of energy in smaller spaces; it can be integrated seamlessly into the greenhouse, without sacrificing growing room. That makes it an excellent fit for small greenhouses that don't have room for water containers.

Though we often compare PCM to passive mass materials, like water, we should highlight that the two work in different ways and have unique pros/cons. Water stores thermal energy by warming up and then slowly radiating heat for long periods, often several days. Phase change materials provide a boost of heat when they undergo a phase change, and are intended to work on a daily basis as the greenhouse temperature fluctuates. If there is a long cold period and the greenhouse does not warm up enough to cross the set temperature, the PCM is not able to function.

FIGURE 12.10. Salt Hydrate PCM. Salt hydrates are another possibility for a PCM product, though less common for buildings. They are cheaper and normally used in larger containers, such as these PVC tubes provided by SavENRG. However, they have shorter lifetimes, due to evaporation of the liquid and corrosiveness of the salts on the packaging material. If they leak, they are more toxic to the growing environment than bio-based alternatives. Credit: SavENRG

Because daily fluctuations are the norm, PCM is a good solution for most greenhouses. It is not appropriate, however, for climates that experience little diurnal fluctuation (like the very hot and humid climates of the southern US).

The main disadvantage with PCM is simply the cost. Compared to water, it's much more expensive per Btu of storage capacity. Thus, you pay for the space-efficiency factor. A good application is to use PCM sparingly and in conjunction with other heat storage materials, not as the sole solution. We typically install it on one or two walls of the greenhouse, which keeps the cost down and gives the greenhouse that added boost of thermal performance. Small urban greenhouses, retrofit projects, passive solar and off-grid greenhouses are all excellent candidates for PCM.

Takeaways

- Water is the cheapest and most effective passive thermal mass material.
- The common rule of thumb is to use 2–5 gallons of water per square foot of glazing area if relying exclusively on water for passive mass/temperature control.
- Any thermal mass material requires diurnal temperature fluctuations to have an effect. You should place mass so that it is directly exposed to light during the heating season. That allows it to store as much thermal energy as possible during the day, to heat the greenhouse at night. Mass can be strategically shaded in the summer.
- Using phase change materials (PCMs) is a good tactic when growing space is limited. They provide a large quantity of heat storage in a very small space, and they can be housed inside a wall instead of taking up floor space in the greenhouse.

Further Reading

Yanda, Bill and Rick Fisher. *The Food and Heat Producing Solar Greenhouse*. John Muir Publications, 1980.

"Passive Solar Design Strategies: Guidelines for Home Building." Download available from the National Renewable Energy Laboratory, nrel.gov/docs/legosti/old/17252.pdf

Chiras, Dan. *The Solar House: Passive Heating and Cooling*. Chelsea Green Publishing, 2002. Many design strategies, with good information on incorporating masonry as thermal mass.

Phase Change Energy Solutions, phasechange.com

Endnotes

1. The British Thermal Unit (Btu) is the standard unit for measuring heating and cooling systems in the US. It is also a direct measure of specific heat, defined as the energy required to raise 1 pound of water 1 degree Fahrenheit.

2. The heat gain of a water barrel depends on its exposure to solar radiation, which varies by greenhouse. 55-gallon drums commonly heat up 5°F–10°F per day if located in the sun. Calculation: 55 gallons × 8.34 Btu/gallon/F × 7°F = 3,211 Btu per drum. 18,200 Btu ÷ 3,211 = 5.6. Roughly 5 or 6 barrels would be required to achieve the same heat storage effect.

Case Study: Cheyenne Botanic Gardens
The Power of Water

4,700 sq. ft. community/educational greenhouse
Cheyenne, Wyoming

The Cheyenne Botanic Gardens is a testament to the power and beauty of water as thermal mass. The center bay of the three-bay greenhouse is lined with turquoise-blue columns which refract and reflect light as they are illuminated by the sun, creating a uniquely serene ambiance. Of course, they are hard at work as well. The water-filled fiberglass columns absorb and store all the energy needed to keep the passive solar greenhouse a minimum of 40°F (4°C) throughout the year, enabling it to showcase fruit trees, ferns and bonsai trees for the public.

The two outer bays of the greenhouse are the production greenhouses, growing vegetables for volunteers and ornamentals for the city of Cheyenne. Here water has a more practical and less attractive variation: the north walls of the side bays are lined with 55-gallon black drums. They keep the greenhouses above freezing through long winters without backup heating.

Now in its 30th year of growing, the Botanic Gardens greenhouse is about to get a new, much larger addition. A four-story conservatory is currently under construction next to the existing greenhouse. The state-of-the art conservatory will feature a tropical forest, event spaces, elevated canopy walkways and educational exhibits. Instead of using polycarbonate (like the existing greenhouse), Director Shane Smith selected a tempered and laminated glass for the whole greenhouse. Though it was a much larger investment, Shane justifies the glass with its longer lifetime and higher insulation value (R-4.2). In his nearly 50 years overseeing the Botanic Gardens' greenhouses, he has had to replace polycarbonate a number of times due to yellowing and hail damage. The new structures include masonry walls as thermal mass, but still hark back to the simple power of water with a pond, a waterfall and interspersed water containers. More information is available at botanic.org.

CHAPTER 13

Using the Earth for Heat Storage

The systems discussed in this chapter all take advantage of the huge source of mass under every greenhouse: the soil underground. Consider that a 12' × 20' greenhouse has 960 cubic feet of soil underneath it, up to a 4' depth. Though soil has a lower volumetric heat capacity than water, the sheer volume makes it an incredibly effective heat sink. It would take roughly 2,000–3,000 gallons of water (over 50 water barrels) to create the same heat storage capacity as this volume of soil.[1] Moreover, the soil stays a stable temperature year-round (discussed in Chapter 6, section titled "Insulating Underground"). Using the Earth as heat storage allows the greenhouse to tap into this huge natural battery. The sun is the heat source, the greenhouse is the solar collector; and the earth is the storage mechanism…it's souped-up thermal storage that relies on natural elements.

These systems are relatively simple, and go by many names. The technical name (what you will find on Wikipedia) is Ground-to-Air Heat Exchanger. We've stuck close to that nomenclature, but given it an easier acronym to say: Ground-to-Air Heat *Transfer* (GAHT) system. Other designers have given it their own names including a Subterranean Heating and Cooling System and Climate Battery. They are related to earth tube systems, though there are some slight variations among them, as described below. For simplicity, we'll stick to "GAHT" for the rest of the book.

Q &A

Is a GAHT system a geothermal heat pump?

No; *geothermal* (meaning heat from the earth) is a broad term that applies to many systems. Often, people jump to the conclusion that a GAHT system is the same as a ground-source heat pump (what are often loosely referred to as "geothermal systems"). While both rely on the stable, warmer temperatures of the soil below the frost line, heat pumps involve compressors and other mechanical equipment that makes them much more complex and expensive. They circulate water or a refrigerant deep underground. When vertically bored, depths are often 150–250 feet below grade. A GAHT system, in contrast, simply circulates air in the shallow earth, roughly 4' below ground. It is a much simpler, much cheaper system that circulates air using fans.

How It Works

GAHT systems serve two vital functions—heating and cooling—depending on the temperature of the greenhouse. When the greenhouse heats up during the day, the GAHT system serves a cooling function. A thermostat turns a fan on that pumps hot air from the greenhouse through a network of pipes underground. The soil absorbs heat from the air, cooling it. After traveling through the pipe network, the cooler, drier air is then exhausted back into the greenhouse.

A primary driver of the cooling process is condensation underground. When hot, humid air is cooled underground, it reaches the dew point, and the water vapor is forced to condense. This further cools the air via the energy transfer in the phase change, or latent heat of condensation. The energy in the humid air is transferred to the droplets of water, and the air is dramatically cooled underground. It is hard to say exactly how much of the GAHT functioning is due to the latent heat of condensation, though we know it is a major factor.

Furthermore, the condensation effect provides some important added benefits: during the day, GAHT systems take humidity out of the air in the greenhouse (where it can lead to pests and disease), force it to condense, and transfer the water into the soil underground (where it can be absorbed by roots). The underground pipes are normally per-

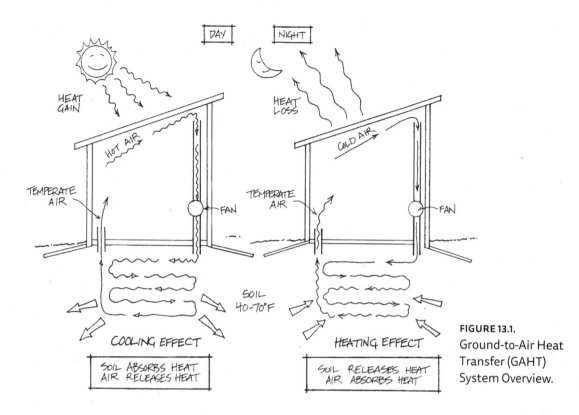

FIGURE 13.1.
Ground-to-Air Heat Transfer (GAHT) System Overview.

forated HDPE drain pipe. Tiny perforations in the pipe allow water to drain out into the soil. A cloth sleeve around the pipe prevents soil from entering and clogging the pipe.

The second function of a GAHT system is heating. Whenever the greenhouse gets cold (at night or in the winter), a second thermostat turns the fan on and begins to extract the heat stored in the soil. The system still works in exactly the same way (air moves in the same direction); the only difference is that now the soil is warmer than the air due to heat loss in the greenhouse at night. During cold periods, the air absorbs heat from the soil as it is circulated underground, providing heat to the greenhouse.

You can see the effect of a GAHT system by looking at the data from a greenhouse in Colorado over a couple of winter days (Fig. 13.2). The GAHT system goes on during two periods: once during the day, and once at night. During the day, you can see that the greenhouse air temperature (shown with the dashed line) rises quickly. When the GAHT turns on, it is stabilized. At night the same effect occurs: the air temperature is falling; the GAHT turns on; and the rate of decline slows. Note how the temperature of the air that is exhausted (the X line) is the same for day and night. This is because there is full heat transfer between the air and the soil; the air is exhausted at about the temperature of the soil. The functioning of the GAHT is the same, but it serves a different purpose depending on the indoor air temperature.

Seasonal Heat Storage

Figure 13.2 shows the GAHT system working over a day; however, it's important to understand that there are seasonal cycles as well. The soil deep underground experiences a moderate fluctuation in the temperature throughout the year. With a GAHT system, the soil beneath the

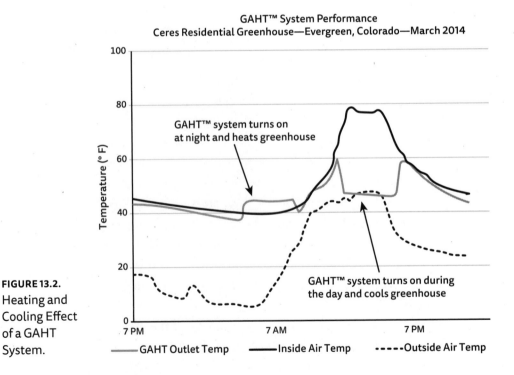

FIGURE 13.2. Heating and Cooling Effect of a GAHT System.

greenhouse still experiences these seasonal fluctuations, but typically it is a few degrees warmer for longer into the year. This is due to the heating effect of the greenhouse.

The analogy of a battery is particularly useful to explain this process. For the warmer half of the year (whenever there are more heat gains than losses), the GAHT system is predominantly heating the soil; it is charging the battery. Compared to external soil temperatures, the soil heats up several degrees up over the summer. In the fall (when the greenhouse needs heat), it starts to draw this stored energy back out, and the soil gradually gets cooler.

At a certain point in the year (usually November/December), the greenhouse maxes out its bank account, so to speak. The soil returns to the natural temperature it would be at that depth outdoors. Fortunately, this is moderate to warm for most climates—about 40°F–55°F (4°C–13°C) year-round—in our location and many others in North America. Thus, the GAHT system can still heat the greenhouse. Though 45°F (7°C) air is not hot, it is much warmer than the nighttime lows outside in the winter in cold climates. To put it another way, the battery (the soil) has two charging sources: hot air from the greenhouse, and the low-grade geothermal energy from the earth. The second allows the GAHT system to provide some heating even during the coldest parts of the year.

A common question is: what is the total heat contribution of the greenhouse to the soil underground? This is hard to say exactly, partly because soil with a warm greenhouse over it will be naturally warmer than the soil outdoors. We know the GAHT system plays a significant role in heating the soil underground, though some natural fluctuations (as occur outdoors deep underground) are still the norm.

Importantly, to better retain stored heat underground, the soil beneath the greenhouse should be insulated. This is the same method used to insulate the perimeter of the greenhouse (discussed in Chapter 6.) Rigid polystyrene insulation can be installed around the perimeter roughly 4' underground when the GAHT system is installed.

History and Context

GAHT systems, and variations of them, have been around for decades. Many of the older solar greenhouse design books show variations of GAHT systems that use a shallow area of rocks as thermal mass. Several research studies on these systems were conducted in the 1970s and 80s; so, clearly, these systems are not new (see Further Reading). They are, however, becoming more popular in greenhouses in the past decade. Much of their advancement is due to the work of the Central Rocky Mountain Permaculture Institute (CRMPI) based in Basalt, Colorado. There, permaculture grower Jerome Osentowski installed similar systems in high-altitude greenhouses. Alongside an engineer, John Cruickshank, and architect, Michael Thompson, they coined the term *Climate Battery*, described in Jerome's book *The Forest Garden Greenhouse*. Climate Batteries use much smaller fans and Jerome does not advocate using underground insulation. Otherwise, the systems are very similar. Climate Batteries have also been called Subterranean Heating and Cooling Systems (SHCS), a name created by the late engineer John Cruickshank.

Another slightly different system is *earth tubes*. These provide one-way airflow into a structure, often an earth-ship home. Earth tubes draw air from the outside, prewarm it by running it through pipes underground, and pump the warmer air into the home. GAHT systems, in contrast, are closed-loop systems: they pump air from the greenhouse underground where it is cooled or warmed, and then exhaust air back into the greenhouse, thus taking advantage of the free heat provided by the greenhouse.

Designing and Installing a GAHT System

Components

Every GAHT system has a few basic components, shown in Fig. 13.3.

- **Intake pipe(s):** An intake pipe extends to the peak of the greenhouse, where the air is hottest. Air is drawn in here before it's circulated underground. Thus, the system circulates the hottest air in the greenhouse, accelerating heat transfer and storing the most energy

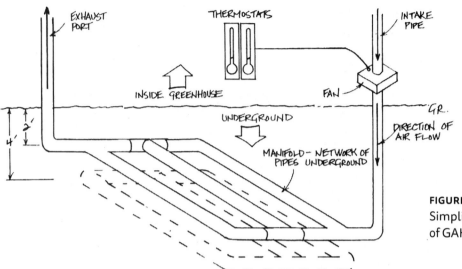

FIGURE 13.3.
Simplified illustration of GAHT components.

in the soil. The intake and exhaust pipes are typically larger than the pipes underground, as they transfer the full volume of air to the underground pipe network.

- **Underground pipe network:** There are a variety of configurations to lay out the network of pipes underground, often called a *manifold* of pipes. Underground pipes are usually made out of HDPE drain pipe, widely available at hardware stores.

- **Exhaust pipe(s):** After air is circulated underground, an exhaust pipe (also called an outlet) returns it to the greenhouse. The exhaust pipes extend to plant level—where you want the moderate temperature air to be.

- **Fan:** An inline fan either pushes air into the system (if located at the intake pipe), or pulls air out (if located at the exhaust pipe). We normally use the "pushing" method, because it is easier to logistically integrate fans into the intake pipes on the tall north wall, rather than the exhaust pipes (often located in or near the growing beds).

- **Thermostats:** Fans are operated by two thermostats, which turn them on when there is a need for heating and a need for cooling. When the greenhouse is a suitable temperature in between, the fan is off.

FIGURE 13.4.
GAHT Thermostats.

Design Considerations

GAHT systems are deceptively simple. A couple of fans, some pipes buried underground... what's so complicated? In actuality, several variables need to be balanced in order for a GAHT system to function properly. If not balanced, it is likely that a system will not effectively heat or cool the air, or not move air at all. Considerations include:

- **Pipe length:** Longer pipes mean the air has longer residence time underground, creating more heat exchange. The drawback is that with longer pipes resistance increases, reducing air flow. There are also diminishing returns to longer lengths; after a point, all the heat has been transferred and extra length doesn't achieve anything more. Most greenhouses (over 12' in length or so) are long enough to create a sufficient length for pipes. Winding or coiling pipes excessively counteracts any benefits.

- **Pressure drop:** When a 1,000 CFM (cubic feet per minute) exhaust fan blows air out of a greenhouse, it does so at roughly 1,000 CFM (given no wind resistance). When a 1,000 CFM fan blows air through the network of pipes underground, the air is exhausted at a much lower flow rate. This occurs because narrow pipes, and their curves and ridges, create resistance on the air, reducing flow. Called *pressure drop* or *pressure loss*, the reduction in flow due to resistance can greatly impact the effectiveness of a GAHT system. In our observations, it is common for fans to exhaust air at 50% of their rated flow. We've seen cases in which the system is not designed properly, and you could not feel any air coming out of the system at all, even with fans were running at full speed. That does very little for temperature regulation, so pressure drop is a critical factor to consider when designing a GAHT system.

There are a number of things you can do to reduce pressure drop. When laying out pipes underground, curves, longer lengths and narrower diameters all increase resistance on air flow. These elements need to be balanced in order to achieve heat transfer underground, while also trying to reduce resistance and create sufficient airflow.[2]

- **Pipe diameter:** Drainpipe comes in a variety of diameters. Larger pipes allow more air to flow through them, and thus more air exchanges; they are suitable in larger greenhouses. Larger pipes also reduce the resistance on the fan, resulting in more airflow. However, larger pipes are much more expensive, and fewer can be used in the same volume; they also result in less surface area being available for air and soil to exchange heat. We primarily use 4" or 6" diameter pipes to for the network of pipes underground, with larger intake and exhaust pipes depending on the pipe configuration.

- **Fan power:** The fans of the system are inline centrifugal fans. The size of the fan (rated in CFM) should be proportional to the diameter and length of the pipes underground. (Ultimately, all these factors depend on the size of the greenhouse and need for climate control.) We generally size fans so that they can make a complete air exchange every 2–5 minutes (so, 12–30 air exchanges per hour). For example, if a greenhouse has 1,000 cubic feet of air, common fan ratings would be 200–500 CFM. (Creating one full exchange every minute would require a 1,000 CFM fan. To create one exchange every 2 minutes, you would need half that flow, or 500 CFM. One exchange every three minutes would need one third that air flow and so on.) We often distribute this load into two separate fans, as we'll go into below.

We find that these are large enough to effectively create enough air exchanges, despite pressure drop. Other designers give more conservative recommendations, only creating five air exchanges per hour. In our experience, due to resistance factors, sizing fans this small can be problematic, though much depends on the size and layout of pipes. An exact number requires an in-depth evaluation of the system design as well as the climate and greenhouse design. Fan sizing depends on length of pipes, and need for air movement in the greenhouse (depending on total heat gain). Thus, the above metrics provide some numbers to start with, but should be modified based on your system. Simple recommendations are hard to come by.

- **Pipe spacing:** A common mistake we see when people build a GAHT system is cramming as many pipes as possible underground, even having them directly touching one another. In order to exchange heat with the soil, each pipe needs to be surrounded by a sufficient mass of soil. We recommend spacing the pipes at least 2' apart. Wider pipes require wider spacing.

- **Even distribution of airflow:** In order for the system to operate effectively, airflow should be evenly distributed throughout the network of pipes underground. If one pipe is shorter than the others, air experiences less resistance through that pipe. As a result, air takes the shortcut. More air travels down the shorter pipe, leading to uneven airflow and heat transfer. The goal is to use the entire mass of soil underground. Accordingly, pipes should all be the same length and size.

- **Depth of pipes:** Up to 4' below grade is a typical depth to access a large volume of soil underground. We normally space pipes out in layers, having one layer 4' below grade and one 2' below grade. One critical caveat: the pipes need to be above the water table at your location. Otherwise they will be flooded with water and inoperable. If you have a water table higher than 4' below grade, this necessitates burying pipes shallower or looking for other heat storage solutions.

As you can tell, there are many variables that determine how a GAHT system functions. It is hard to synthesize all of these into a simple equation. There are certainly *complex* formulas that simulate changes in each variable, but these are impractical for most readers. Instead, we recommend relying on examples of successful installations, the recommendations above, and using professional advice or engineering to tailor a GAHT system to your greenhouse and climate.

System Design

There are a couple of variations to laying out and installing the GAHT system pipes underground. Another term for the series of pipes is a *manifold*, because just like the manifold in your car, a GAHT system involves one large pipe (called the intake pipe, or sometimes the *riser*) that brings air underground to a series of smaller pipes. This allows air to be distributed evenly through a large volume of soil.

The first approach, shown in Fig. 13.5, uses a large intake pipe that brings air to a series of smaller pipes laid out straight (or nearly straight) across the greenhouse footprint. These are connected to another large exhaust pipe on the other side.

This design uses evenly spaced layers of pipes for the manifold. One layer is buried roughly 2' below grade, the other about 4' below grade. (The exact depths do not matter so long as there is sufficient soil surrounding each layer.) Both layers can connect to one large intake pipe, as shown in Fig. 13.8. Alternatively, the layers can be separated into their own independent systems. In that case, each layer has its own fan and intake and exhaust ports (shown in Fig. 13.5). When layers are separated,

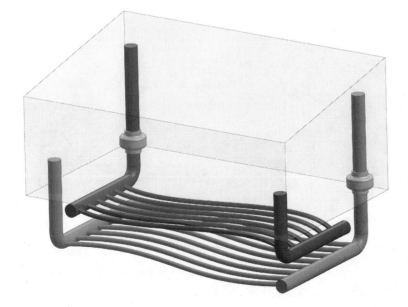

FIGURE 13.5.
Layered
Configuration
of a GAHT
System.

it allows you to distribute the air flow—that means two smaller fans, and smaller intake/exhaust pipes—which can reduce the need to source very large, expensive pipes.

The second approach we call the *barrel approach*. A centrally located barrel serves as the intake pipe. The smaller pipes are drilled into the barrel and extend radially from this point. The barrel is often a low-cost 55-gallon PVC drum, which saves the cost of purchasing large-diameter drain pipe.

The downside with this approach is that it creates an irregular pattern of pipes underground: some pipes have to curve and wind more to cross the greenhouse; others are nearly straight. Those with more turns encounter more resistance, which slows airflow, leading to uneven heat distribution through the mass of soil underground. Best practice dictates that all pipes in the pipe network should be the same length so, we don't typically recommend this approach.

Installation

Installing a GAHT system can be divided into two parts. First, the pipes are buried underground before the greenhouse is built. After the greenhouse is constructed, the fans and thermostats are installed in the

FIGURE 13.6.
The pipes in Rob Torcellini's greenhouse are connected to a rocket mass heater (not a GAHT system), but still exemplify the barrel approach to system design. Credit: Bigelow Brook Farm

greenhouse. Below is an overview of the basic steps, which may need to be modified for your specific application. For brevity, we paint these processes in broad strokes. For more detail, consult a professional.

Installing Pipes Underground

1. **Excavate a pit underground using a backhoe or excavator.**
2. **Install pipe network underground.** This can either be done directly in the excavated pit, or (for small systems) separately on a level surface beforehand. When positioned in the pit, the intake and outlet pipes should extend above grade (ground level), as shown in Fig. 13.7. These will connect to the fans inside the greenhouse.

   Pipes are connected via adapters (connection pieces like a T-joint) or by directly drilling and inserting one pipe into the other, as shown in Fig. 13.8.
3. **Backfill.** Carefully backfill the removed soil, preventing any large rocks from crushing the pipes in the process. This involves one or two workers in the pit to help distribute soil around the pipes, as shown in Fig. 13.9.

FIGURE 13.7.

FIGURE 13.8.

FIGURE 13.9.
Credit: Driftless
HomeWrights

4. **Compact the soil.** After the pit is backfilled, the soil should be allowed to settle before building the greenhouse. This can happen naturally, using water and time, or by using a compactor during backfilling. Once the soil is firm and level with grade, the greenhouse can be built. Keep in mind that the intake and exhaust pipes will extend above grade. These need to be positioned according to your greenhouse floor plan so that they can be connected to the fans inside the greenhouse.

What is the best backfill material?

We normally use the native soil as backfill material because it is the most practical and cost-effective (i.e., free). From a performance perspective, a fine-grade gravel is slightly better. It has a higher heat capacity and higher conductivity (rate of heat transfer) than most soils. Sand or large stones should not be used. (Sand has a low rate of conduction, and large stones create air pockets.)

Whether to upgrade to gravel is simply a matter of cost. Filling the entire pit with gravel will get quite expensive, and the payoff is uncertain. A good middle road is to use a small amount of gravel directly surrounding the pipes, and backfill the areas in between with the original soil. Again, we stick with simple soil, which has good performance, mostly because it can hold a lot of water (which has the highest heat capacity). You also don't have to deal with removing a huge mound of soil after the install.

Above-ground Installation

Installing the fans, thermostats and pipes inside the greenhouse is a separate process that happens after construction. If you don't have electrical experience, we recommend hiring an electrician to complete these steps.

1. **Install intake pipes and fans.** An intake pipe should extend to the peak of the greenhouse; it can be made of PVC or HDPE drain pipe. We normally install the inline fans in the intake pipes so they push air through the pipes. Fans are wall mounted, as shown in Fig. 13.10.
2. **Install exhaust pipes.** Exhaust pipes should extend to plant level in order to distribute temperate air near the plants. We normally locate these in the corners, where they are out of the way.

FIGURE 13.10. Intake and GAHT Fan in a Small Greenhouse. A commercial-size version is shown in the color section: the Climate Battery in The GrowHaus.

3. **Install thermostats.** An important part of the system function is setting up proper thermostatic controls. Two thermostats control each fan (as shown in Fig. 13.4). One turns the fan on when the greenhouse gets too cold; the other when it gets too hot. When the greenhouse is a suitable temperature in between, the fan is off. This avoids excessive energy usage, because the fan isn't running all the time.

Performance

One way to evaluate the effectiveness of a GAHT system is by looking at the temperature difference between the inlet and outlet. Comparing the temperature of the air when it went into the system versus when it is exhausted out tells you how much heat was transferred to the soil. Ideally, the air should achieve "full heat transfer": all of the heat is absorbed or released and the air is exhausted at roughly same temperature as the soil. We find this is achievable for most of the year.

To better understand how a GAHT system performs, the first question is: what is the temperature of

Fan Controls

The set temperatures depend on your growing goals and ideal indoor temperature; 40°F–90°F (4°C–32°C) is a common target range. Keep in mind that the GAHT system does not respond immediately; it takes time to create enough air exchanges and control temperature. Thus, we recommend setting the "heating" thermostat a bit warmer than your actual low threshold (e.g., 50°F [10°C]) to give it some time to stabilize the falling temperature. Likewise, 75°F (24°C) would be a logical set temperature for the "cooling" thermostat if your actual threshold is

90°F. Labels, as shown in Fig. 13.4, are helpful to remember which thermostat is which and their set temperatures.

Because this can get confusing, at Ceres we provide wiring diagrams to explain the concept and installation to electricians. Though simple when you think about it, using two thermostats and two set temperatures is atypical for fan controls (which usually only serve cooling). If hiring an electrician, it is crucial to clearly explain this using diagrams and clear instructions to avoid mishaps.

the soil? With an insulated foundation, the soil beneath the greenhouse is generally 40°F–60°F (4°C–16°C; averaged across the US), meaning a properly functioning GAHT will exhaust air in the same range. Though 40°F is not *hot* air, during the winter, it can keep the greenhouse from freezing—using very little energy.

How much heating and cooling does a GAHT system provide over-all? Several confounding variables like the soil type, soil temperatures, design of the system, humidity levels and heat gain make it hard to give a simple answer to this question. One study evaluated 18 commercial greenhouses in Europe that had various applications of ground-to-air heat exchangers.[3] Though extremely varied in their design and location, most of these greenhouses experienced energy savings in the range of 40%–60%.

We've drawn similar conclusions from observing many greenhouses with GAHT systems: they typically provide the bulk of the heating and cooling load. Like passive thermal mass, they are not perfect climate control systems. Usually a cooling method, like ventilation, is necessary to supplement the cooling ability of the GAHT during the day. Most growers we know also have a backup heating system. This is usually re-quired on extremely cold nights in the winter, when the GAHT's un-derground battery may max out. It can still provide heating, exhausting air at the temperature of the soil deep underground, but this is often insufficient if trying to keep the greenhouse above a higher threshold.

If you do integrate a backup heater, we always recommend turning the GAHT system *off* when the heater is *on*. GAHT systems draw heat down into the soil. That is ideal during the day when the greenhouse provides the heat, but results in wasteful energy use if a fossil-fuel-based heater is providing the heat. In other words, you want the heater to heat the greenhouse air, not the soil. The two systems should not be running at the same time.

To disable the heating function of a GAHT system, you can set the heating thermostat (the one with the low set temperature) very low (e.g., 30°F [–1°C]). The backup heater will be set at a higher temperature than this, so it will engage first and should keep the greenhouse well above the GAHT fan set point. The GAHT fan is still plugged in and can still run during the day to provide cooling, but it cannot run at night because the greenhouse will not reach the low temperature threshold.

Similarly, the set temperatures of exhaust fans should be several degrees higher than the GAHT fan. This allows the GAHT fan some time to run, storing heat in the soil before the exhaust fan turns on. They may sometimes be on simultaneously—meaning the GAHT pumps heat into the ground and the exhaust fan exhausts it outside, though this is not a huge issue because neither fan is very energy intensive, like an electric heater.

Takeaways

- GAHT systems provide both heating and cooling by storing excess heat of the greenhouse during the day in the soil underground.
- There are different variations for the layout of the pipes underground. We recommend using multiple layers of pipes laid out linearly.
- A number of variables go into designing a high-performance GAHT system, mainly finding the right balance between pipe length, pipe size and fan power in order to achieve good air flow rates and heat transfer underground.
- While GAHT systems are do-it-yourself systems for many homeowners, they often require additional instructions, and should be customized to suit your greenhouse.

Further Reading/Resources
Frequently Asked Questions, system design/consulting/instructions: ceresgs.com

Osentowski, Jerome. *The Forest Garden Greenhouse*. Chelsea Green Publishing, 2015. Gives good information about Climate Battery systems.

Though significantly different from GAHT systems, resources explaining understanding earth tubes can be helpful as context. See, for example: *Passive Annual Heat Storage: Improving the Design of Earth Shelters*, by John Hait. Rocky Mountain Research Center, revised edition, 2013.

Endnotes

1. Assumptions: volumetric heat capacity of damp soil is 25 Btu/ft^3/F°. Water has a heat capacity of 8.34 Btu/gallon/F°. Calculation: 960 × 25 = 24,000 Btu/ft^3/F°. 24,000/8.34 = 2,877 gallons of water, or roughly 52 barrels.

2. On a technical note, some resistance is needed because it affects the type of air movement, making it turbulent rather than laminar. Turbulent flow is needed to achieve heat transfer. Most systems naturally achieve turbulent flow due to the great number of curves and ridges in the pipes.

3. M. Santamouris, et al., "Use of Buried Pipes for Energy Conservation in Cooling of Agricultural Greenhouses," *Solar Energy* vol. 55, No. 2, pp. 111–124, 1995.

CHAPTER 14

Solar Hot Water

Commonly used for residential water heating, solar hot water systems are very effective heating systems. They can also be used in greenhouses, though they are less common. They are particularly strategic if you have a need for water heating, such as heating a fish tank in aquaponic systems. In those cases, solar hot water systems can provide both water and space heating for the greenhouse.

The drawback is that solar hot water systems do not provide any cooling. Unlike GAHT systems, there is no thermal storage; the system does not remove heat from the greenhouse. Additionally, solar hot water is more expensive than the systems discussed thus far. Components like collectors, pumps, water storage tanks and controls require a significant upfront investment, even for small systems. A basic residential system would start in the range of $5,000–$10,000, installed. Thus, they are best suited for mid-size to large greenhouses, typically commercial operations or schools, where that investment can be quickly recouped due to economies of scale.

System Design

Fig. 14.1 shows a typical configuration for a drain-back solar hot water system. In a drain-back system, a pump circulates water through a collector mounted on the roof or next to the greenhouse. The water heats up as it moves through the collector, and returns to a hot water storage tank, moving through a heat exchanger in the tank. (Variations exist for housing the heat exchanger in a drain-back tank instead.) Hot water is stored in the tank. From there, another thermostatically controlled pump moves the hot water into the greenhouse.

Drain-back systems circulate water as the fluid. When the pump turns off, water drains out of the collector into a drain-back tank. Without this feature, water would stay in the collector and pipes and would freeze on cold nights, breaking the components.

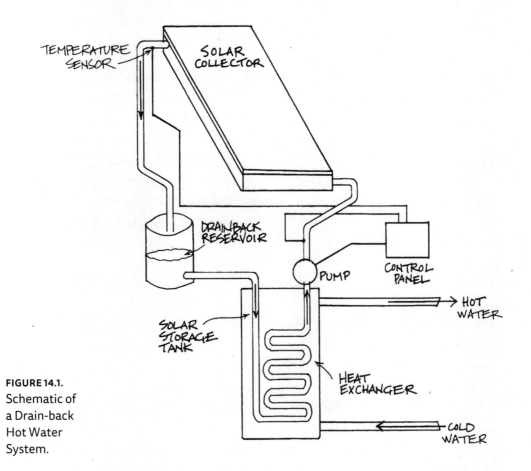

FIGURE 14.1.
Schematic of a Drain-back Hot Water System.

An alternative to a drain-back system is to use propylene glycol, a non-toxic chemical, as a circulating fluid. Glycol systems don't need a drain-back tank since the fluid doesn't freeze. However, the fluid is more expensive and needs replacing. If considering a solar hot water system, talk to a professional installer to decide whether a drain-back or glycol system would be most effective for your situation.

Collectors

There are two types of solar hot water collectors: flat plate and evacuated tube collectors. Flat plate collectors use simple components. Water circulates through copper pipes on a flat plate painted with a highly absorbent coating. Water simply heats up as it moves through the hot coils.

Evacuated tubes are newer technology that adds a process to increase the efficiency of the collector. Each evacuated tube contains a heat pipe—a closed vacuum with a fluid inside. The low-pressure of the sealed environment allows liquid to evaporate at lower temperatures. Heat pipes rely on the heat transfer of phase changes (described in Chapter 12): as liquid evaporates, it absorbs a huge amount of energy. The vapor is then forced to condense to water again in the pipe, and as it does so, it then releases that energy. This heat is collected in a heat exchanger at the top of each tube. The heat exchanger circulates a fluid that then moves to the hot water tank.

The mechanics of evacuated tubes can be complicated, but the takeaway is that they increase the efficiency of the collector by using the physics of phase changes. This allows them to produce more heat, particularly in cold and cloudy climates. A relatively new technology, the economics and payback of evacuated tubes are still emerging. We recommend getting a variety of quotes and opinions on the different options from solar installers in your area.

A professional installer should help you evaluate the placement and orientation of the collector. Like solar panels, collectors should be angled and oriented to maximize heat absorption during the seasons when hot water is needed.

Storage Tanks and Delivering Heat to the Greenhouse

After water is heated in the collector, it needs to be stored somewhere so that it can heat the greenhouse when needed. Conventionally, this is supplied by a hot water tank. The tank can be housed in the greenhouse or in a separate insulated storage area. This leads to one of the disadvantages of solar hot water systems: tanks take up room in the greenhouse.

Case Study: R&H Unlimited Greenhouse
Three Ways to Use Solar Energy

384 sq. ft. prototype greenhouse
Leonard, Michigan

The demonstration greenhouse designed by R&H Unlimited, a small design and engineering firm in Michigan, uses passive solar design. It is also heated with a solar hot water system and powered by solar photovoltaic panels, exemplifying the several ways solar energy can be harnessed in a greenhouse.

The solar hot water system consists of two 4' × 8' evacuated tube collectors housed above the roof. They transfer hot water to a 1,000-gallon tank on the north side of the greenhouse. The hot water is delivered to the greenhouse via radiant floor heating (coils laid in sand in the floor).

Additionally, the greenhouse integrates a small, 540 watt, solar photovoltaic system to power the pumps and energy-efficient LED lighting to supplement lighting in the winter. Combined, the multiple systems allowed the greenhouse to grow abundantly year-round and completely off-grid, producing fresh ripe tomatoes through Michigan's cold and cloudy winters. However, multiple renewable energy systems created a high price tag. In total, R&H Unlimited spent about $30,000 to build the demonstration greenhouse. At over $200 per sq. ft., it proves that the hurdles to year-round growing in net-zero-energy greenhouses are not technological. It is simply a matter of budgets and getting there cost-effectively.

FIGURE 14.2. Credit: R&H Unlimited

However, there are some creative work-arounds. For one, tanks can be built underground. Billy Mann at Sagebrush Solar, for instance, has installed concrete tanks underground in his greenhouse applications in Sun Valley, Idaho. (Mann also designed the system in the Sage School greenhouse, described in the case study at end of this chapter.) Built into the foundation, the tanks store hot water without sacrificing growing space in the greenhouse.

Another alternative is to eliminate the storage tank altogether and circulate hot water directly through the greenhouse. Specifically, the hot water can circulate through PEX tubing in a concrete floor (radiant floor heating). The water heats the slab during the day, and the slab slowly radiates that heat at night. The downside is that the slab will also be passively heating the greenhouse during the day, when it heat is not needed. The ideal storage mechanism should be able to deliver heat on demand. Another variation is to heat the water for fish tanks in an aquaponics growing system, though this involves more complicated controls.

Generally, controls consist of two thermostatically controlled pumps that move water through the system. One pump monitors the temperature of the water tank and compares it to the temperature in the collectors. When hot water is available, the pump turns on and replenishes the tank. A second pump monitors the temperature of the greenhouse. When it gets too cold, it circulates hot water from the tank into the greenhouse.

Inside the greenhouse, hot water is typically circulated through PEX tubing, either in a concrete slab or in/under growing beds. Bed heating is more effective because it delivers heat directly to where it's needed—the soil—rather than heating the entire air space.

If you add an aquaponics system into the equation, it can necessitate a third pump to monitor the water temperature in the fish tank. Fish have different temperature requirements than plants or the air, so the system needs to be controlled independently. All these pumps and controls clarify why solar hot water systems are more expensive than the other options discussed thus far. They require professional installation and can necessitate complicated controls.

Case Study: The Sage School, Idaho Net-Zero-Energy Aquaponics Greenhouse

2,400 sq. ft. educational greenhouse
Sun Valley, Idaho

When The Sage School of Sun Valley, Idaho, moved into a new building, it inherited an inefficient polycarbonate greenhouse. Head of School Harry Weekes immediately started applying for grants to renovate the structure and turn it into a functional and sustainable year-round learning environment.

With $75,000 in grant money and individual donations, The Sage School renovated the structure in 2012, adding a solar hot water system and a large aquaponics growing system. Today, the greenhouse serves as a living classroom and provides supplemental income for the school, which sells the produce to area restaurants and at the local farmer's market during the summer.

Four 4' × 8' evacuated tube solar collectors provide 100% of the greenhouse's heating needs. The installer, Sagebrush Solar, added insulated water tanks inside the greenhouse. From there, hot water is circulated through coils in the growing beds, keeping the root zone at roughly 60°F (15°C) year-round. That reduces the need to heat the entire air space in the greenhouse. "The greens are fine even when it is below freezing outside because the soil stays 60°F," according to greenhouse manager Sara Berman.

The solar hot water system also heats the aquaponics fish tanks, typically the #1 energy consumer in an aquaponics greenhouse. Heating tanks in addition to the greenhouse introduced

FIGURE 14.3. Credit: The Sage School

a new challenge: how not to boil the fish. Tilapia have a narrow temperature range that they thrive in, and they cannot tolerate water above 95°F (35°C). The water coming out of the hot water system is 200°F (93°C), on average. In order to not overheat the fish, designer Billy Mann installed an additional thermostatic control that operates the heat exchanger for the fish tanks.

There are two strategies for a fish tank heat exchanger. You can either circulate the hot water tank water through a coil that sits inside the fish tank, or vice versa, circulate the fish tank water through a coil inside the hot water tank. Mann decided to go with the latter, since having a coil inside the fish tank would disrupt the natural behavior of the fish. This creates the

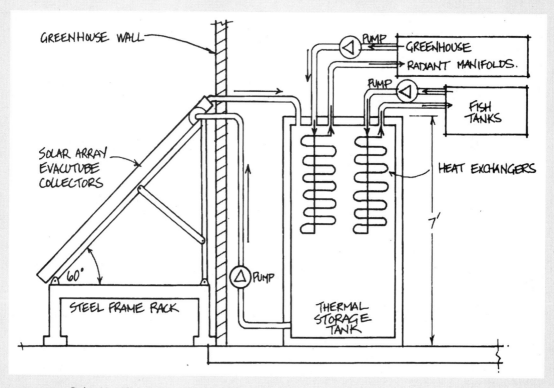

FIGURE 14.4. Solar Hot Water Heating System, Integrated with Aquaponics. Credit: Billy Mann, Sagebrush Solar

possibility that sludge from the fish tank water (which contains waste particles) can build up inside the coil. Mann addresses that problem by flushing the coils every couple of years with a high pressure hose. An overview of the setup is shown in Fig. 14.4.

The greenhouse is also equipped with a backup propane heater, but since the renovation, it hasn't been needed. The solar hot water system provides all of the heat for both plants and fish, saving the school about $2,800 in propane heating costs every year. Head of School, Harry Weekes, sums it up:

> "We managed to keep the greenhouse above 40 degrees from November through March, and never turned on the heat. We grew lettuce, radishes, kale, arugula, and spinach from December through March. The only reason we pulled these out was to plant over 300 tomato plants, which are now over five feet tall and headed to the rafters—literally. During our first crop of tomatoes last year, one woman stopped by to see what kind of trees we were growing. Suffice it to say, the system is up and functioning well.

> "In the last three to four months, we have sold over 200 pounds of produce, and made more than $4,000 from our sales. As we launch into our second season of tomatoes, and start our peppers, cucumbers, basil, and nasturtium, we anticipate the greenhouse sustaining itself economically within another 12 months (not to mention that it is producing food the entire time!).

> "We have found a way to reduce our energy costs, generate income, produce local food, and to reduce our environmental impact. In short, the students have seen the practical, possible, and very tangible ways they can make a difference in these areas."

CHAPTER 15

Rocket Mass Stoves and Compost Heaters

The final heating systems we'll discuss use biomass (either wood or compostable material) to heat the greenhouse. Like solar hot water, these are purely heating systems; there is no cooling effect during the day.

Biomass systems have some unique advantages. They produce very high temperatures and deliver heat quickly to the greenhouse. Moreover, they don't rely on solar energy or heat from the greenhouse. That makes them a good option for greenhouses in climates with harsh and low-light winters. Many of the examples in this chapter are in the northeastern or midwestern US—places with harsh, cloudy winters.

The downside of both of these systems is that they are more time intensive to maintain or operate. Rocket mass stoves need to be lit manually. Compost piles need to be rebuilt every couple of years and occasionally aerated, strenuous and time-consuming work. Though more hands-on, either can work well in the right applications. This chapter explores how to integrate them with a greenhouse. Like many topics, they warrant much more in-depth research if you want to build a system yourself. See Further Reading at the end of this chapter.

Rocket Mass Heaters

A rocket mass heater combines a highly efficient wood stove (called a *rocket stove*) with thermal mass, usually made of cob (a mixture of clay, sand and straw or other fiber). The stove burns wood at extremely high temperatures, heating up the surrounding mass, which then slowly radiates that heat over the course of the night or cloudy day. In short, rocket mass stoves rely on wood's ability to give off tremendous amounts of heat and thermal mass' ability to store that heat and release it slowly. They are becoming increasingly popular as an efficient way to heat both homes and greenhouses.

How It Works

Fig. 15.1 shows the basic design of a rocket mass heater. The stove part of a rocket mass heater is made out of an insulated combustion chamber and a *heat riser*: a tall, vertical column that draws smoke out, creating an intense chimney effect. The combustion chamber is horizontal, with wood fed in vertically. The fire, in other words, burns sideways. This feature allows the smoke and flames to run horizontally before hitting a 90-degree turn at the start of the heat riser. The turn creates turbulence, causing the smoke and air to mix before it goes up the riser.

The riser gets extremely hot, causing secondary combustion—the burning of gases and particulates in the smoke. This makes rocket mass

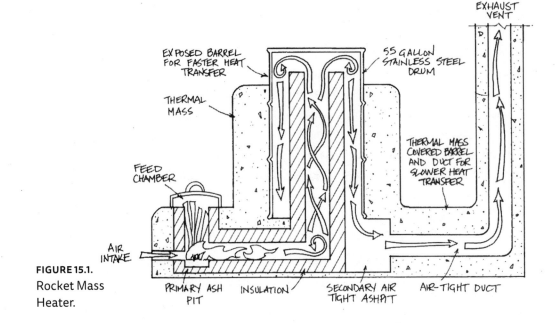

FIGURE 15.1.
Rocket Mass
Heater.

heaters immensely more efficient compared to traditional wood stoves. In effect, they use much less wood to heat a space.

The riser also creates a strong "chimney effect"—a natural convection current that draws air up and out of the stove. This strong draft draws and moves air into the stove. It also allows the stove to burn without smoke/embers drafting back into the greenhouse, making it safe to operate without constant attention. The riser's draft-effect is what gives the rocket heater its name: stand by one, and you can hear the "whooshing" of air being drawn up quickly through the heater.

The final leg of a rocket mass heater is an exhaust flue, which carries the smoke out of the greenhouse. (Though we call it smoke, the exhaust from a rocket mass heater is actually mostly water and CO_2. It has little smell because the particulates that give regular smoke its odor and dark color are burned off in the heat riser.)

Thermal mass, usually in the form of cob (a mixture of sand, straw and clay that's cheap and easily molded) surrounds most of the stove, riser and exhaust flue. The cob absorbs the thousands of Btus produced

FIGURE 15.2.
Rocket Mass Heater.
Credit: Verge Permaculture

by the stove and by the smoke as it's carried out of the structure. Often, cob surrounds the exhaust flue and forms a bench, running along or around the edges of the greenhouse, as shown in Fig. 15.2. This allows it to absorb any remaining heat before it's exhausted outside. The exhaust flue exits the greenhouse out of the top of the structure. It should extend well above the roofline to avoid downdrafts.

Rocket mass heaters are normally DIY systems, made of low-cost components. It is crucial to use the right materials—those that can withstand the extremely high temperatures. Instead of standard brick, for instance, you must use fire brick. Additionally, consider the greenhouse's humidity when building your system. If using cob, it should be protected with a plaster. If considering a rocket mass heater, we recommend that you do thorough research on how to build a system (see Further Reading) and/or take a workshop.

Pros, Cons and Best Applications
Pros of a Rocket Mass Heater

- **Efficient:** The stove uses far less wood than a conventional wood stove, meaning less wood to stock and keep near the greenhouse. It also utilizes the heat much more efficiently, by slowly heating up the mass of the stove.
- **Long-lasting heat:** After the stove has burned for 3–5 hours, the thermal mass should be warmed enough to provide heat for the rest of the night, or longer (depending on the size of the stove and mass). That long-lasting high-grade heat is useful in climates with very harsh winters where GAHT systems may not provide enough heat.
- **Cheap:** The stove is usually made out of recycled materials (brick, steel pipe or barrels, and cob). Most mass stoves typically cost less than $300 in materials.
- **Added thermal mass:** Even when the stove is not lit, the thermal mass of the heater helps regulate the greenhouse temperature to a small extent.

- **Multi-purpose:** A warm bench can be useful for seed trays or a cold backside. Mass heaters can also be combined with hot tubs, aquaponics tanks and other integrated systems.

Cons

- **Time and maintenance:** The biggest drawback is that rocket mass stoves require an active operator to keep the stove burning for a few hours while it heats up the mass. According to Rob Avis of Verge Permaculture (see case study Chapter 3). A wood-fed stove needs to be lit for a few hours in order to heat up the mass. During that time it needs to be refueled every 15 to 30 minutes. In Rob's view, that makes them "great in a house where you can sit by the fire with a book, but it's an inconvenience when you have to run out to the greenhouse." If you have a detached greenhouse and don't enjoy the idea of trudging through snow on winter night to stoke a fire, this is not a good system for you.

"Rocket Mass Heater on Steroids"

That is the title of a video clip explaining Rob Torcellini's prototype rocket mass heater in his dome greenhouse in Connecticut (see case study at the end of Chapter 8). Rob built a pellet feeder to supply the rocket mass heater with pellets at a rate of 12 lbs. of pellets per hour. The feeder uses gravity to feed the fire at a constant rate, allowing for longer burn times without having to refuel or kindle the fire. The heater is partially buried underground, and heat is drawn through exhaust pipes buried in a layer of sand. In this way, the heater takes advantage of the existing thermal mass underground, rather than occupying space in the greenhouse. Rob continues to make adjustments to his heater, still in its testing phase, though it has been effectively heating the greenhouse and aquaponic fish tanks. "I was able to maintain the building temperature at 60°F [16°C] degrees (30°F outside) [–1°C], and heated the water from 55°F–63°F [13°C–17°C]," says Rob. For a full explanation of the system, you can see videos on the YouTube page of Bigelow Brook Farm: youtube.com/user/web4deb.

Variations exist to make rocket mass heaters run longer and be easier to operate. In cold climates, people are building larger "batch box"-style heaters that burn larger batches of wood for longer.

- **Self-built:** Not necessarily a "con," rocket mass stoves are almost always self-designed and built. Building a stove is not difficult, once you know what to do, but they do require time, research and initiative. According to Leslie Jackson, co-author of *Rocket Mass Heaters:* "From designing, building, processing firewood for maximizing burning efficiency, troubleshooting, maintaining, and living with, this is not a hands-off heater.... It's one of the reasons people love it."

- **Feed stocks required:** As wood-burning stoves, rocket mass heaters require a stockpile of wood in or around the greenhouse. Another common modification is to add a grate so the stove can burn wood pellets. If these are cheaply sourced in your area, using pellets means you avoid having to chop, dry and store wood; it also allows for easier feeding in some cases.

- **Space:** Normally, the thermal mass of a rocket mass heater takes up significant space in a greenhouse that could otherwise be used for growing.

- **Increased drafts:** Not necessarily a detriment, but something to be aware of, is that rocket mass heaters increase air infiltration when they are burning. The stove is intaking air from the greenhouse and exhausting it outside. This, in turn, draws cold outdoor air into the greenhouse through leaks and cracks. Since the heater is simultaneously warming the air, the effect is not that detrimental. Some growers may prefer it as it helps reduce the humidity, replacing the existing air with fresh, albeit very cold, air.

Due to all these pros and cons, we find rocket mass heaters to be very useful as backup heaters if you have the room for the thermal mass, and don't mind operating a stove when it's needed. They can supplement another heating method during very cold periods. However, relying on them constantly as a heating source throughout the colder months can get cumbersome, both in time and materials (firewood).

Compost Heat Recovery

The final method of this section is less common, but equally viable if done correctly in large-scale applications. Compost heating takes advantage of the fact that a large compost pile produces millions of Btus of heat per day, reaching 110°F–160°F (43°C–71°C) at its center. That is a lot of free, sustainably generated heat. Additionally, at the end of a compost pile's life, it leaves a high-quality organic soil amendment/fertilizer.

The challenge with composting is simply the volume of material that has to be moved and stored. To heat a residential greenhouse, compost piles are typically 10'–20' in diameter and 8'–10' tall—this is not your backyard compost bin. Smaller compost piles don't get hot enough to be effective heating elements. Namely, a compost pile must get hot enough (occasionally over 150°F [65°C]) to kill the eggs of parasites, cysts and flies. While a large compost pile is a heating machine, a small compost pile is simply a pile of rotting organic material…not something you want around your greenhouse.

Thus, compost heat recovery is best done at scale, using large mounds of compost that can generate very high temperatures. They are

FIGURE 15.3.
Compost Piles with Hydronic Heating Coils. Credit: Gaelan Brown

best suited for growers who can easily source and move big volumes of compostable material (i.e., those who have machinery and composting material on-site).

If you find compost heating bizarre, recall from Chapter 2 that for most of greenhouse history this was a primary method of heating greenhouses. Today, compost systems can be as advanced as solar hot water or GAHT systems. That said, they are still less common, mostly because they require moving large amounts of organic material and are labor intensive to create and maintain.

How It Works

The magic ingredient in a compost pile is the aerobic bacteria that produce heat as they consume and break down organic material. Given the right inputs—organic material, air and water—this process occurs naturally, generating an incredible amount of heat completely on its own.

The key to success of a high-heat compost pile is the right balance of inputs: carbon, nitrogen, air and water. All of these parts are needed, in the right proportion. That balancing act makes compost heating more time intensive than other heating systems. It requires research, tinkering and possibly a few tries to get right.

Building a High-heat Compost Pile

Compost piles should have a carbon-to-nitrogen ratio (C:N) between 20:1 and 30:1. In other words, for every part nitrogen there should be 20–30 parts carbon. Each material has its own C:N makeup. Most compost materials, like leaves and wood chips, are predominantly carbon. Materials like manure and coffee grounds are predominantly used to add nitrogen. (Fun fact: hair clippings have extremely high levels of nitrogen. If you have a longhaired friend who is ready for a change, you can give your compost a boost of nitrogen that way.) Online "compost calculators" can help you create a pile with the right C:N ratio based on your inputs. Or you can research others good substrates, usually a combination of organic matter, like wood chips, and manure.

Because aerobic bacteria require oxygen, compost piles need to be aerated. Using a substrate material that is not too dense, like wood chips or mulch allows both air and bacteria to contact the material. Air, not organic material, is often the limiting factor when it comes to heat production.

A compost pile will produce heat for a few months to a couple years, depending on pile size, composition and aeration. After the bacteria have consumed all the organic material, the temperature will slowly decline, and the pile will need to be reconstructed with new organic matter.

Outdoor piles should be insulated with a layer of wood chips or straw to reduce heat loss from the center of the pile during colder months. If the pile is built correctly, the center of the pile should stay at 110°F–160°F (43°C–71°C). There are a few different methods to draw this heat into the greenhouse, ranging from very simple (direct use) to very advanced (steam heat exchangers).

Small-scale and Residential Applications

On the simple end, some growers use compost directly inside or outside the greenhouse. Will Allen's urban farm in Milwaukee, Wisconsin, for example, heats several hoop houses with rows of compost piled alongside the structures (see case study, Chapter 4). The compost directly heats the greenhouse through the thin polyethylene plastic. This is only a logical means of heating for uninsulated greenhouses (like polyethylene hoop houses) which have terrible energy efficiency. While effective, this is a labor-intensive method. Growing Power processes millions of pounds of compost every year; work crews and many volunteers are needed to move it around.

A more common method for residential growers is using the compost inside the greenhouse by creating *hot beds.* In this case, the bottom of a raised bed is filled with manure and organic material. As the compost degrades, it delivers heat to the plant roots. The top 1'–2' of the bed is filled with well-aerated garden soil. This acts as a bio-filter, absorbing toxic byproducts—ammonia and excess nitrogen—produced by raw compost. It also reduces the risk of attracting insects, though pathogens

FIGURE 15.4.
Compost Pile at Growing Power. Mounds of compost are stored on site before being piled around the hoop houses. Credit: Growing Power

are still a risk. Because compost piles produce odors, as well as ammonia, they should not be directly exposed to the greenhouse air.

A hot bed will heat the soil for several months, after which it loses its heating capacity, and the bed must be rebuilt. This is back-breaking work. For that reason, presumably, we haven't seen many examples of hot beds, but they do work for some people. Roger Marshall, author of *How to Build Your Own Greenhouse*, describes creating a hot bed to overwinter plants in his New England greenhouse. He filled the bottom 3' of his raised bed with a mixture of horse manure, grass clippings, wood chips and leaves. Marshall notes that it took 4–6 weeks for the bed to heat up to maximum temperature (160°F [71°C]). "It will remain at this temperature for about 2 months before gradually cooling down to about 80°F [27°C], warm enough to keep plants growing for a long time after they've stopped growing outdoors."[1]

Using compost directly is simple but labor intensive. If hand shoveling manure and composting materials is not your cup of tea, this is not a good fit for you. Direct use of compost also introduces the risk of pathogens and insects. A properly functioning compost pile should kill anything harmful, but "properly functioning" is the operative phrase. If the pile does not get hot enough (over 150°F [67°C]), it will be a hotbed mainly for insects. For those reasons alone, we don't recommend direct use of compost unless you are completely committed to it.

Compost Water Heaters

A more advanced method of using compost is combining it with a heat exchanger. Hydronic heat exchangers circulate water through the compost pile, and then transfer this heat to the greenhouse. This method was popularized by Jean Pain, a French inventor who spent many years developing systems on his farm in Vermont. Pain used compost water heaters to heat his home and several outbuildings.

Compost water heaters are relatively simple systems. PEX tubing is built into the center of a large compost pile, as shown in Fig. 15.3. A water pump circulates water through the tubing, where it's exposed to the 100°F–160°F (38°C–71°C) temperatures at the center of the pile. The hot water is then pumped to a hot water tank inside the greenhouse, where it's stored until needed for heating. Additional PEX tubing pumps the hot water from the tank through the greenhouse, either through the floor (as radiant floor heating) or under seed trays, delivering heat to plant starts.

Compost water heaters allow you to extract much more heat from the pile by circulating water through the hot core of the pile. They also avoid having to store compost next to the greenhouse. The pile can be kept in an adjacent area, and insulated PEX tubing connects the two. Finally, hydronic systems are controllable: a thermostat inside the greenhouse draws hot water from the water tank when it's needed. Unlike direct use systems, the pile does not heat the greenhouse when heat isn't needed (on warm days). Since the output is hot water, this system can also have other uses when not needed for greenhouse heating, such as an outdoor shower or hot tub.

Compost water heaters require physical labor, usually from a few people. According to the Compost Power Network, a compost pile can be constructed in a day by three to five people. It is best done with some machinery and a tractor, though hand shoveling is always required since compost must be carefully layered around the coils.

Perhaps the biggest drawback, in our opinion, is that the pile must be rebuilt every few months to a couple of years. Whereas a GAHT system or a solar hot water system works for many years with little to no maintenance, a compost pile must be reconstructed or moved frequently. However, for farms that have existing compostable wastes, a composting system can make use of a waste stream that would otherwise go to a landfill, saving money and leaving nutrient-rich compost as a byproduct.

Case Study: Ben Falk Greenhouse
Is Small-scale Compost Heating Practical?

266 sq. ft. residential greenhouse
Mad River Valley, Vermont

To add to his homestead in northern Vermont, Ben Falk, author of *The Resilient Farm and Homestead*, built an attached greenhouse onto the south side of his barn. The greenhouse was originally heated with a compost water heater. In the first season, the compost pile proved the heating potential of composting: the hot water was temporarily used for a hot tub and hot showers. However, after a few months, the heat ran out due to a lack of aeration. Ben moved the pile so that it could get more air, and it produced heat for another couple months before the heat slowly trickled to a stop.

"People aren't able to get them to work like Jean Pain did," is Ben's honest review. "The likelihood of failure is high. It's appropriate at the right scale, where you have a lot of manure and the equipment to move it, but not for most greenhouses." Currently, Ben Falk uses his greenhouse for three-season production. He no longer uses compost heat, but does have a wood stove in the greenhouse as an "insurance policy" for cold snaps.

Other growers, like Alice and Karel Starek of The Golden Hoof Farm in Boulder, Colorado, (see case study, Chapter 16) are able to get longer run times and are enthusiastic about compost's heating potential. "Our piles have been lasting about 8 months. So far, it's the chickens that have been the demise of the piles. We use mainly wood chips and have a coil of 4" perforated drain tile at the bottom and up through the center to get plenty of aeration" says Alice. The Stareks use their 12' diameter mound to heat a 300 sq. ft. insulated barn, and they find that it is an appropriate size for the application.

FIGURE 15.5. Credit: Ben Falk/Whole Systems Design, LLC

Takeaways

- Both rocket mass heaters and most compost piles produce high-temperature heat that can be used regardless of outdoor weather conditions. Thus, they are good options for cold climates that have little solar gain in the winter.
- Building a compost pile is labor intensive and requires large amounts of compostable material and the space to store it, making these systems more appropriate for greenhouses in rural areas and impractical for urban backyards.
- Rocket mass heaters must be manually operated to provide heating.

Further Reading

Rocket Mass Heaters
Evans, Ianto and Leslie Jackson. *Rocket Mass Heaters*.
Permies.com web forum.

Composting
Brown, Gaelan. *The Compost Powered Water Heater*. Countryman, 2014.
The Compost Power Network, compostpower.org
Compost mixture spreadsheets from Cornell University: compost.css.cornell .edu/download.html
"Compost Fundamentals," Washington State University, whatcom.wsu.edu/ag /compost/fundamentals

Endnotes

1. Marshall, Roger. *How to Build Your Own Greenhouse*. Storey Publishing, p. 70.

Case Study: Jasper Hill Farm
The Green Machine: Converting Waste to Energy

2,300 sq. ft. commercial greenhouse
Greensboro Bend, Vermont

Located in Greensboro, Vermont, Jasper Hill Farm is a picturesque dairy and cheese farm that heats its 2,300 sq. ft. greenhouse entirely with compost. Forty-seven grass-fed Ayrshire cows are the hub of the farm. Not only do they produce the milk for Jasper's award-winning cheese, but ultimately, they produce most of the energy for the farm.

The secret to this ultra-sustainable farm is "The Green Machine," a multi-functional manure processing system that integrates three technologies to process and extract heat from manure. A bio-filtration system separates solids to produce useable irrigation water. An anaerobic digesters converts solid waste to methane to fuel the farm's boiler. Third, a compost heat recovery system converts the manure into heat for the large greenhouse.

The compost heat recovery system is an advanced design that takes advantage of the

FIGURE 15.6. Inside the Jasper Hill Farm greenhouse, large pipes form part of an advanced heat exchanger. (An exterior view appears in the color section.)
Credit: Jasper Hill Farm/Bob Montgomery Images

steam produced by large compost piles. Designed by Agrilab Technologies, the system uses fans to draw steam downward and pumps it through pipes buried underground. A heat exchanger then extracts heat as steam undergoes a phase change to water. The energy is transferred to hot water and hot air is delivered to the greenhouse through radiant floor heating.

Jasper Hill Farm's compost heat exchanger system is much more complex, intended for farms with more than 100 cows or equivalent feedstocks. Though a much higher capital investment, it also has a much larger energy production, capable of keeping the greenhouse "balmy" year-round, according to a farm manager. Currently, the greenhouse grows lemons, bananas, tomatoes and fig trees year-round in their boreal climate zone. Though the Jasper Hill system was funded by a state grant, it has the potential to pay for itself in five years, according to Gaelan Brown, author of "The Compost Power Water Heater," which describes the system more in-depth.

More than energy production, The Green Machine solves the large problem of pollution caused by waste runoff. Conventional manure management practices often equate to "manure lagoons," which threaten water and air quality in addition to producing immense carbon emissions.

With The Green Machine, Jasper Hill is able to upcycle a former waste product into clean water and energy for the farm. Though still in its infancy, Jasper Hill sees huge potential for large-scale compost heat recovery, particularly on a community scale. "Say you have a livestock operation on one side of the road and a commercial greenhouse production on the other. The output of one could be the heating resource for the other." In that way, large-scale compost heat recovery tackles a multitude of issues, enabling sustainable food production, safe manure management, and carbon reduction. More information about The Green Machine and Jasper Hill Farm can be found at agrilabtech.com and jasperhillfarm.com.

FIGURE 15.7 Bringing It All Together: Sustainable Heating and Cooling Systems

System	Summary	Pros	Cons	Best Applications
Thermal Mass	Dense materials to passively store energy for heating and cooling	• Low-cost • No electricity usage • DIY installation	• Occupies a large amount of space in the greenhouse • Depends on sufficient solar gain to provide heating • Limited control	Residential greenhouses that do not require electricity (passive solar greenhouses), particularly those in sunny climates. Not good for climates with cloudy winters
Phase Change Material	Space-efficient material that stores energy via a phase change from liquid to solid	• Moderate cost • No electricity usage • DIY installation	• Depends on sufficient temperature fluctuations in the greenhouse	Residential greenhouses with limited space; passive solar greenhouses and off-grid greenhouses
GAHT Systems	Stores excess heat in the greenhouse during the day underground using a system of fans and buried pipes	• Provides long-term heat storage (heat stored in the summer can be used in the winter) • Automated by thermostat control • Greater capacity to store heat than passive thermal mass	• Moderate to high cost • Requires excavation underneath the greenhouse (longer installation time) • Requires fans, which use some electricity and create noise	A wide range of greenhouses (both large and small) that use electricity
Solar Hot Water	A common system used in homes, water is heated in panels and used for heating the greenhouse	• Provides high-grade heat for most or all of the year in sunny climates	• High cost • Requires a more complicated control system and usually professional installation • Only provides heating, not cooling	• Larger greenhouses in sunny climates; • Greenhouses that have another need for water heating, such as aquaponic greenhouses • School and commercial operations
Rocket Mass Heaters	Combines an efficient wood stove and thermal mass	• Low-cost • DIY installation • Provides high-grade heat regardless of the outdoor weather	• No automation • Requires an active operator to keep the stove burning while it heats up the thermal mass • Requires a supply of wood or pellets and space for mass in the greenhouse	Residential greenhouses in climates with harsh and cloudy winters; excellent as a potential backup heating solution
Compost Heating	A variety of methods using the heat of decomposing organic material (hot beds and compost water heaters)	• Provides high-grade heat regardless of the outdoor weather; automated • Other benefits: use of compost at the end of the process, sustainable use of a waste material	• Requires large feed stocks of organic matter (and space to store it) • Labor intensive (compost piles must be built and rebuilt)	Large greenhouses in rural areas with lots of compostable material, particularly those in harsh climates with cloudy winters

Backup Heating

The systems mentioned thus far are designed to heat the greenhouse for the majority of the year. However, they are not all perfect heating solutions. Trying to cover the last 5% of the heating requirements during the coldest days of the year often requires a backup heater to meet certain temperature thresholds.

If the greenhouse is small and only needs heating during cold snaps, a simple electric space heater, like you find at hardware stores, can suffice. They are energy-intensive devices and can quickly get expensive if used for too large a space.

Larger greenhouses can turn to propane or natural gas heaters, whichever fuel is more economical in your area. Propane has a slightly higher heat output. They can be sized for a range of greenhouse sizes, both small and large.

Finally, wood-burning stoves are a more sustainable option. Unlike the methods above, these must be manually operated, which is difficult to do through a long cold night and introduces the risk of fire. For these reasons, we recommend using a rocket mass heater as a safer and more effective alternative.

To size a backup heater, you can perform a simple heat loss calculation using an online heat loss calculator like the one provided at builditsolar. com. First you must estimate the greatest possible temperature difference the greenhouse could experience, based on the coldest expected temperature and the target indoor temperature. You can then estimate the heat loss from the greenhouse under these conditions. A heater should be sized to accommodate for the heat loss at that period. You can also use product spec sheets to help size a heater.

In any situation, you can reduce the heat requirement by delivering heat directly to where it is needed. For instance, plants can withstand greater fluctuations in air temperatures if their roots are kept warm. Thus, a wise tactic is to use electric or hydronic heat directly beneath seed trays or growing racks, keeping the soil warm without trying to heat the entire air space. If only your aquaponics fish tanks require heat, heat the water directly, not the entire greenhouse.

CHAPTER 16

Powering the Greenhouse

Many of the heating/cooling systems we've discussed involve electrical components to move and transfer sustainably generated heat. An inline fan in a GAHT system, a water pump in a solar hot water system, exhaust fans, or simply an overhead light…these are small electric loads, but they do require electricity. Getting that electricity to the greenhouse—conventionally or renewably—is the focus of this chapter.

The first question to ask is: does the greenhouse *need* power? It is possible to grow year-round in many climates without electrical components, creating a passive solar greenhouse. If electricity is needed, most growers connect the greenhouse to the grid. There's also the option of adding solar photovoltaic (PV) panels to the greenhouse to produce electricity renewably. Adding solar panels allows a greenhouse to produce as much or more energy than it consumes, becoming a net-zero-energy, or beyond-net-zero-energy structure. Similar to homes, the options of integrating solar panels on a greenhouse include connecting to the grid, adding battery backup or creating a completely off-grid system.

No Power: Passive Solar Greenhouses

Creating a completely passive greenhouse is an attractive option for many residential growers, and understandably so. It saves the initial cost of bringing electricity to the structure. Moreover, it avoids ongoing electricity costs, though these are usually minimal for residential greenhouses. Finally, passive greenhouses are quieter, more relaxing spaces because they don't involve running electric fans.

Common Electrical Systems in a Greenhouse

- Exhaust fan
- Circulation fan
- GAHT fans
- Overhead lighting
- Water pumps for irrigation

To clarify terms, we use *passive solar greenhouse* to describe a greenhouse that has *no* electrical components. Passive solar greenhouses are inherently off-grid. It's also possible to have an *off-grid* greenhouse that uses electricity but is powered by a stand-alone solar PV system with battery backup. Both are independent of the grid, but they are very different greenhouses.

The primary drawback with a passive structure is that it offers less control over the growing environment. If there is a record-breaking cold spell or hot streak (not unusual occurrences these days), adding a backup heater or fan is not an easy option. For that reason, passive solar greenhouses are at risk of greater temperature fluctuations. They typically freeze at some point in climates with long freezing periods. Good air circulation can be difficult to achieve in the winter because there are no circulation fans. That makes them more prone to issues related to excessive humidity in the winter (when vents need to be closed to keep out cold air).

Given these pros and cons, passive solar greenhouses are a good fit for growers who don't need a tightly controlled environment. They are often used as three-season greenhouses, shutting down over the most extreme months when temperature control is more difficult. They are not a good fit for commercial greenhouses that have money and profit on the line and depend on the greenhouse maintaining more controlled conditions.

Recommended Systems for Passive Solar Greenhouses

- **Ventilation:** Passive ventilation is critical since this is the greenhouse's source of cooling and indoor air circulation. Don't skimp on vent area. Some vents won't be needed during the winter; these can be insulated and sealed. Full venting is necessary to control overheating and provide good air circulation in the spring, summer and fall.

Solar-powered fans can provide supplemental venting. These use electricity, but a small built-in PV panel powers them so they don't require wiring or electrical hookup. We have found these to be good options to easily add mechanical ventilation to an otherwise completely passive greenhouse.

- **Heating:** Consider passive thermal mass materials, like water and/or phase change material (see Chapter 12). These help store heat during the day and deliver it at night. A rocket mass heater or wood stove (see Chapter 15) can be used for backup heating.
- **Cooling:** In addition to venting, shade cloth helps prevent overheating in the summer.
- **Lighting:** Though not absolutely necessary, an overhead light is a nice addition to the greenhouse in case you want to check on things at night. In passive greenhouses, standard bulbs can be replaced with battery-operated LED lights.

Conventional Power: Connecting to the Grid

Most residential growers wire the greenhouse directly to the home or a nearby structure if they want power. The greenhouse then runs off standard power from the grid. The size and complexity of wiring a greenhouse varies greatly by the energy demand. If the greenhouse occasionally requires a backup heater, but does not need power full time, a simple extension cord from the home can suffice. On the other end of the spectrum, commercial greenhouses with large electric loads need to contact the local utility to evaluate the best way to connect to the grid.

If you don't have experience doing electrical wiring, we recommend hiring an electrician to complete this process. If you plan on wiring the greenhouse yourself, we still recommend having your work checked by an electrician to verify your own and the greenhouse's safety. Below, we provide an overview of wiring a greenhouse to the home.

The process is like wiring any detached structure or addition, which you or an electrician can complete. First, determine all the appliances you will be running and the wattage of each. Additional circuits are added to the home's main breaker panel based on the total power draw (wattage) of all components running in the greenhouse. If you divide the total *wattage* by the *voltage* (usually 120) you get the *amperage* needed (from the equation watts = amps × volts). From there, an electrician can determine how many circuits are needed and the amperage of each. We usually install one 20-amp circuit for residential greenhouses. If you will be using grow lights or other energy-intensive equipment, the greenhouse will need several circuits (another reason to evaluate energy-efficient lights like LEDs).

Typically, electric cable housed in conduit is buried underground and run from the home's breaker panel to the greenhouse, as shown in Fig. 16.1. If wiring the greenhouse yourself, consult the many resources available for creating a wiring diagram and doing the calculations. Roger Marshall spends a chapter in *How to Build Your Own Greenhouse* on this topic.

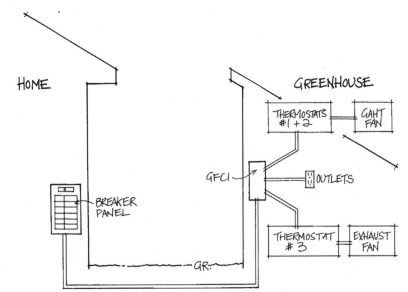

FIGURE 16.1.
Wiring a Greenhouse to the Home.

Tips on Wiring a Greenhouse

- Given the high-humidity of a greenhouse, we recommend adding a ground-fault circuit interrupter (GFCI) where the cable (coming from the breaker panel) enters the greenhouse, as shown in Fig. 16.1. A GFCI will shut off power if there is a ground fault or contact with moisture. That way, if any water comes into contact with a socket, the power shuts off rather than shocking you.
- All outlets should have weatherproof covers, as shown in Fig. 16.2. This prevents water (either from condensation or watering the garden) getting into the outlet and creating a short circuit.
- Consider what equipment you will be using and factor that into the placement of outlets. We usually put outlets on the side walls of the greenhouse near the doors and corners. That leaves the longer north and south wall open for growing.

FIGURE 16.2.
Covered Outlets.

Integrating Solar PV

A common misconception is that "solar greenhouse" refers to "a greenhouse with solar panels." Indeed, solar photovoltaic (PV) panels are what people most associate with the word "solar." But there's a reason these two don't always match up. While solar panels can be a source of renewable electricity for the greenhouse, they are not a good source of direct heating.

Converting electricity into hot air (called *resistive heating*) is extremely energy intensive compared to the methods described in Chapters 12–15. Instead of using the sun's energy directly, resistive heaters powered with solar panels require inefficient conversions: First, solar panels convert sunlight into electricity, then, a heater converts the electricity to hot air. A much wiser use of the sun's energy is to use the hot air already present in the greenhouse during the day and store it via thermal mass or another method. Solar panels *are* very useful for powering components that facilitate thermal storage, like fans or pumps.

There are several ways to integrate solar panels into the greenhouse. To find the right one for you, first clarify your objectives. Are you concerned about the reliability of power at your location? Is there no access to the grid? Or, do you simply want to offset carbon emissions from the greenhouse's electricity usage?

There is no wrong reason to invest in solar panels; but it is important to be clear about your goals, so you can find the best way to meet them. For example, some residential clients tell us they want to add solar panels to their backyard greenhouse so that "it's sustainable." We ask them if they already have solar PV on their house. "No, we haven't thought of that," they say. This is perplexing. If your goal is to reduce carbon emissions, why not start with the home, where solar panels can have a much greater impact?

PV systems have significant economies of scale because certain components are needed regardless of size. Adding a larger system on your home allows you to get the greatest return on your investment and offset the most carbon emissions. Even though homeowners say they "want a sustainable, solar-powered greenhouse," what they really want is to *live* more sustainably. Clarifying that fact allows them to site a PV system where it can do the most good.

Other growers have valid reasons for adding PV panels to the greenhouse. The top of a large south-facing greenhouse roof can be a good site for a PV system (as long as it does not detract from the glazing area too much). Some locations have frequent power outages; others don't have power access. Finally, some growers are morally driven to be independent from the electric grid. Each of these situations lends itself to a particular PV strategy.

The options for integrating solar panels are the same as for homes: You can add a grid-tied system (with or without battery backup), or you can go off-grid with a stand-alone system that includes a large battery bank. If the greenhouse has a very small electrical load, you may also have the option of purchasing a small kit system and running the greenhouse entirely in DC. Each strategy has pros and cons, summarized in Fig. 16.3.

FIGURE 16.3: Solar Photovoltaic Systems—Overview

System	Size	Pros	Cons	Best if . . .
Kits	Very small (< 500 W)	Very cheap; DIY installation	Small power production; typically small voltage (12V) that can only be used with certain appliances	The greenhouse has just a couple appliances (very small loads)
Grid-tied	>1 kW	Cheapest method to install a significant-size system. Uses electricity most efficiently. No maintenance	Does not protect against power outages	The primary motivation is reducing carbon emissions and there is access to the electric grid
Grid-tied with Battery Backup	>1 kW	Provides reliable power and is cheaper than an off-grid system	More expensive than a simple grid-tied system	The greenhouse must have reliable power and there is access to the electric grid
Off-grid (Stand-alone)	Any. Very large systems get extremely expensive	Provides independence from the electric grid, and access to power anywhere	Very expensive; high maintenance	The greenhouse does not have access to the electric grid (very rural areas) or the primary motivation is to be independent from the grid

Across all of them, PV panels can be housed on the greenhouse roof or ground-mounted next to it. In small greenhouses, roof placements can cover a significant amount of glazing, shading the greenhouse. If possible, use existing south-facing roof space on adjacent structures (like a house). It is likely that there is more space, which can be better utilized.

Small-scale (Kit) Systems

You may have seen a portable solar panel kits at a hardware store or from online retailers. These are becoming increasingly popular for charging small appliances like phones, often used at camping sites and other remote applications. However, many are large enough that they can power a greenhouse with a small electric demand.

Kit systems are similar to stand-alone/off-grid systems, except the power capacity of these systems is typically 500 W (watts) or less. In comparison, a mid-size residential PV system would be in the range of 5,000 W (5 kilowatts [kW]), ten times the size. Thus, they are good if the electric demand is less than 2 kWh per day, which is about the energy expenditure of a single circulation fan. (For more on how to determine energy usage, see "Sizing a Greenhouse PV System" below.) Most kit systems come with the restriction that they provide power in low voltages, like 12 V. Thus, all appliances in the greenhouse must be 12 V, which limits your options for equipment (since most things run in 120 V).

If applicable, kit systems are low cost, typically less than $1,000. They are commonly self-installed, which takes out much of the cost of solar. Kits include an inverter (to convert DC power to AC for appliances). They can also be hooked up to a small 12-volt battery.

Completely DC

Solar panels produce direct current (DC) electricity. Batteries, likewise, store and discharge DC. However, most appliances run on alternating current (AC). Converting DC to AC requires an inverter, which happens to be one of the most expensive parts of a PV system.

Inventive DIYers can save money by avoiding an inverter and running the greenhouse in DC. Power can be stored in low-cost batteries like 12 V car batteries to reduce costs even further. There are also efficiency gains (between 10%–30% depending on end-use equipment) when you do not have to convert DC to AC. The practicality of this all depends on the electrical loads and whether it's possible to find the electrical components you need (fans, pumps, etc.) in direct current.

Though options are more limited for DC appliances, they do exist, thanks largely to the RV and boat industries. Snap-Fan is a company that makes DC exhaust fans specifically for off-grid greenhouses. The solar attic fans mentioned in Chapter 7 also apply, since they are designed to run directly off the panel. If you have few appliances and are installing your own system, you can evaluate whether to run the whole greenhouse in DC.

Grid-tied Systems

Grid-tied systems currently comprise the majority of PV applications, as they are the simplest and cheapest way to add a decent-size solar PV system. In a grid-tied system, the solar panels feed power to the home. They are also directly connected to the grid so that the grid can absorb or supplement power if the panels produce more or less than the greenhouse needs.

Most utilities offer "net metering," which gives you credit for any surplus power produced. If panels produce more energy than the greenhouse needs, the surplus is fed back to the grid, causing the electric meter to run backward. When the greenhouse has a higher power demand than the panels can supply, the grid supplements power.

Net metering allows the grid to act like an infinite battery—it absorbs the excess energy when it is plentiful, and fills in when more is needed. Furthermore, as a storage mechanism, the grid is much more efficient than batteries. Storing energy in physical batteries creates conversion losses. The electrical energy produced by the panels must be converted to chemical energy stored in the batteries, and this is not 100% efficient. In other words, you can't get 100% of the energy produced back when using batteries. With grid-tied systems, the electricity simply slips on and off the grid in the same form. You get credit for all of the electrons produced.

Grid-tied systems are also smaller than off-grid, because the system does not have to produce all the power needed for each day. If there is a cloudy day, and the panels can't produce enough, the grid supplements. On a sunny day, the solar panels "make up" for those electrons. Grid-tied systems are sized to meet the average annual demand, rather than the daily demand. In contrast, off-grid systems need to be large enough to supply power for the extremes—the rare week of cold and cloudy weather. That makes off-grid systems larger and more expensive.

The downside to grid-tied systems is that they don't protect against power outages. If there is a power failure, the system can't operate. If power outages are a concern in your area, and consistent power is essential for your greenhouse, you will need to add batteries for backup power.

Grid-tied with Battery Backup

Grid-tied with battery backup systems are the middle ground between grid-tied and off-grid systems. These function exactly like a grid-tied system 99% of the time. The 1% difference is in the case of a power outage. If the grid goes down, a small battery bank takes over and powers the greenhouse. Some advanced components make the switch from grid power to battery power automatic and seamless.

Batteries and added controls make battery backup systems significantly more expensive than grid-tied systems. Thus, they only make sense for growers who absolutely need consistent power. They are cheaper, however, than off-grid systems. Thus, the battery bank is sized to only power "critical loads" during a power outage. The batteries in a hybrid system are only about one third the size of an off-grid system.

Critical loads are the components considered essential to keep running. In a greenhouse, critical loads may include parts of an aquaponics growing system, like water pumps or aerators, or climate controls, if growing high-value crops like medical marijuana. The majority of greenhouses, however, can survive a power outage without much consequence. For that reason, grid-tied with battery backup systems are rare in greenhouses. Most growers that invest in solar PV either opt for grid-tied or go completely off-grid.

Going Off-grid

Note: Much of the information in this section was provided in Dan Chiras's course on off-grid solar (see Further Reading), which we recommend for a more detailed exploration of how to design and install an off-grid system.

We've already mentioned small-scale off-grid systems—those that can power a fan or a light but not much else. Full-blown off-grid systems involve much larger PV systems and large battery banks capable of powering the greenhouse every day of the year. For some growers, off-grid systems fulfill a romantic ideal of severing the tie to the utility and living independently. For others, they are a practical necessity if there is no utility connection at the greenhouse site. In either case, it's important to evaluate the realities and challenges of going off-grid before diving in.

An off-grid solar PV greenhouse is both the power consumer and the power plant. All the electricity the greenhouse needs is produced *and stored* on-site. But what if there is week without sun? Hopefully, you have a large enough battery bank to tide the greenhouse over. This requirement—large battery storage—puts off-grid systems in class of their own in terms of cost.

Additionally, the PV array in an off-grid system must be sized much larger than a grid-tied version because the panels must power the electric loads and also keep the batteries charged. The size needed depends on the climate—specifically, the average length of a typical overcast period. In sunny Colorado, for example, a battery system is commonly sized to provide power for three consecutive days. In the Midwest, the norm is five days, given the cloudier climate.

As you can tell, the cost and practicality of an off-grid PV system depends on your climate. Cloudy climates require much larger battery banks and PV arrays, which greatly increase the cost. The reasonableness of going off-grid also strongly depends on the power demand of the greenhouse. The size of the electric load determines how big of a "power plant" you need to build. If you are trying to run numerous HID (high intensity discharge) grow lights for 16 hours a day, costs will quickly go through the roof. If you are only trying to power a couple small electronic devices that run infrequently, adding a small battery bank is doable.

Invest in Efficiency First

By creating a more energy-efficient greenhouse, you reduce the total energy demand. In turn, you reduce the size of the PV system you need to install. Buying the most efficient fans, lights, pumps, etc. saves you money in the end because you avoid the cost of installing additional solar panels.

When it comes to efficiency, we recommend doing everything possible to avoid having to run electric heaters. Also called resistive heaters, these are incredibly energy-intensive equipment. A small electric space heater uses 1,500 watts. A standard CFL light bulb, in comparison, is 12–14 watts. Investing more in insulation and passive thermal storage will pay off hugely in the end because you won't have to power heaters. A rocket mass heater or nonelectric system can provide backup heating during intense cold periods. The same applies for grow lights: upgrading to the highest efficiency LEDs is a no-brainer if going off-grid. In short, you can invest the money in either efficiency or a larger battery system. The better investment is always efficiency.

The Battery Bank

The two most common battery types for off-grid solar applications are lead-acid and lithium-ion. Lead-acid batteries (specifically, flooded lead-acid, also called *wet cell*) are the traditional choice for the large battery banks required of off-grid systems. They are the cheapest way to store lots of energy. The downside is that they require significant maintenance. The liquid solution in the battery must be checked monthly.

Lead-acid batteries are intended to deep-cycle, but they should not remain overly discharged for long periods. This reduces their lifespan. If properly cared for, deep-cycle lead-acid batteries can last 10–15 years, but many first-time owners fry the batteries in just 4–5 years because they underestimate the need for maintenance. If you kill the battery bank, you'll need to make the huge initial investment all over again. Finally, lead-acid batteries should be housed in a protected environment where they will not get wet, damaged, freeze or overheat.

The alternative is lithium-ion batteries, which are making inroads into large-scale power storage. Currently more expensive than lead-acid, their costs are declining rapidly—to the point where they will soon be cost-competitive per watt with lead-acid. Lithium-ion batteries come with many practical advantages. First, they don't have the maintenance and safety considerations of lead-acid batteries. They also have high power densities—they're able to store more energy in a smaller volume. The Tesla Powerwall, for instance, made by the electric car company, is a compact lithium-ion battery that can store about 6.4 kWh of electricity. Currently, it costs \$3,000 for the battery plus about \$2,000 for the inverter. Adding installation, the cost likely comes to over \$6,000. That is expensive energy storage compared to lead-acid batteries, but many people are drawn to its sleek, compact form. It can be housed on a wall instead of requiring a separate room and is virtually maintenance-free.

Whatever battery option you go for, it's important to realize that large-scale battery storage is likely a few thousand dollar investment. That is on top of the cost of the solar panels, inverter, and other necessary components if using standard AC equipment. In total, this puts the cost of typical off-grid system for a residential greenhouse over \$10,000. We point that out solely to provide a "reality check" for potential off-grid growers: though it is a very enticing option, building a decent-size off-grid system only makes sense if you are strongly motivated to be independent from the electric grid, or if you do not have access to the grid at your site. (The latter case justifies off-grid PV because hooking up to the electric utility will probably require bringing power lines out to the site—an even larger cost.)

The good news is that costs for solar panels and battery systems are dropping quickly due to mass production. Whatever size or scale of system you are considering, consult a PV installer about your options. The economics of PV vary hugely by area. Only by evaluating the economics of solar panels in your location (given productivity/peak sun-hours, tax incentives and local utility rates) can you understand the cost-benefit of going off-grid.

FIGURE 16.4.
Tesla Powerwall.
Credit: Tesla

Sizing a Greenhouse PV System

The size of a solar PV system depends on the power demand of the equipment in the greenhouse, how often the equipment runs, and the energy available from the sun in a particular site (peak sun-hours per day). You have control over the first two variables, but the third depends on your location. Additionally, design factors like the tilt and orientation of the panels, shading and panel efficiency affect the energy output, and thus the size of the system.

The Difference between Watts and Kilowatt Hours (kWh)

Watts is a *rate* of energy usage. Think of it like the speed a car is traveling. Watt-hours is a unit of energy, derived by multiplying the rate (watts) by the time the electrical component is running. Think of it as the total distance a car travels. 1 kilowatt-hour (kWh) = 1,000 watt-hours.

While we can't flesh out every detail of sizing a PV system (that should be done in conjunction with a professional or more resources), we can give a basic overview of the process in order to help elucidate whether it is in the realm of budgetary possibility. The basic steps are:

1. First you must determine what electrical components will be running in the greenhouse. Find the power usage (given in watts) for each component. Then, estimate the number of hours each component will be running per day. Multiply this by the watts of that component to determine the total energy usage; the result is in kilowatt hours (kWh).

2. Add up the energy demand of each component to determine the total energy demand of the greenhouse on a monthly or annual basis.

3. The next step is more complicated. You must now figure out what size PV system is required to produce that much energy at your location. The easiest way to approach this is to use online tools that evaluate the solar resource at your location. A website like PVWatts (pvwatts.nrel.gov) can estimate the power production of a PV system at your site. Since you don't know the size of your system yet, PVWatts requires narrowing down the range with estimates. Start with a 1 kW system, and see the annual energy production in kWh. Then adjust to get to your energy demand. In other online calculators, you can input the energy demand and get a system size, but they are not as accurate in terms of climate data. You can also use rule-of-thumb metrics for the average annual energy production of a solar panel in your area. These can be found online or by talking to solar PV installers.

Using this basic process, you can get a ballpark estimate of the required system size. Though the cost of solar PV is consistently declining, it is worth noting that as of this writing the cost per installed watt is $3–$4 in the US for a grid-tied system. Keep in mind that this range does not include batteries or any components needed to create a battery backup or off-grid system.

The examples below walk you through this process using three hypothetical greenhouses and their equipment. The goal is to compare the system sizes and note how they change as electric loads are added. We assume that the solar panels in these examples are located in Denver, Colorado (a relatively sunny location).

Example 1: Small Backyard Greenhouse with Just One Fan

In this scenario, the owners rely on passive systems for climate control. They install water barrels and phase change material. They also use battery-operated lights and passive venting to reduce the electric demand to nearly zero. They only want to run a circulation fan, and thus need a very small system.

- Components: 1 circulation fan (40 watts)
- Energy consumption per day: 120 watt-hours (runs for 3 hours per day × 40 watts)
- Average energy consumption per year: 44 kWh (120 watt-hours × 365 days)
- Can be supplied by a 100-watt PV system (commonly, a kit system)

Example 2: Residential Greenhouse with Vent Fan and GAHT System

In this scenario, the growers have opted for a 200-watt exhaust fan in addition to a circulation fan. They also use a GAHT system to keep their larger greenhouse at a more stable temperature throughout the year.

- Components: 1 exhaust fan (200 watts), 2 inline GAHT fans (each 120 watts)
- Energy consumption per day: 2.8 kWh

- Assuming exhaust fan runs for 2 hours per day and GAHT fans run for 10 hours per day: (200 watts × 2 hours) + (120 watts × 10 hours × 2 fans)
- Average energy consumption per year: 1,022 kWh
- Can be supplied by a 0.63 kW PV system

Example 3: Small-scale Aquaponics Greenhouse

In our final example, the growers in the previous example decide to add an aquaponics system. They live in a cold climate and will need to incor-

Case Study: The Golden Hoof Farm
The Off-grid Tropical Greenhouse

3,000 sq. ft. commercial greenhouse
Boulder, Colorado

Step into the solar greenhouse at The Golden Hoof Farm outside Boulder, Colorado, and you're greeted not by a monoculture of greens, but by a forest of fruit trees—guava, avocado, mango, banana, lemon, dragon fruit and cinnamon, to name a few. Taller trees create a canopy above, while shorter shade-tolerant varieties form the understory, mimicking the layers of a tropical forest. The floor of the greenhouse is covered with more conventional crops, like kale, chard, mustard greens and tomatoes. In between the vegetables, a green blanket of nitrogen-fixing clover borders pathways and fills in any excess spaces. On a spring day, you step from the dry Colorado landscape, surrounded by corn fields, into this lush humid forest, and it's clear that this is anything but your ordinary commercial greenhouse.

The greenhouse is a small component of Alice and Karel Starek's dynamic "slow-food" farm: they produce sustainably raised pork, lamb, beef, chicken, duck, eggs and vegetables. The birds are fed pre-consumer kitchen scraps from local restaurants and roam freely around the acres of property. Pigs are also fed pre-consumer waste, while the ruminants rotationally graze the pastures. All in all, The Golden Hoof operates more like a farm in the

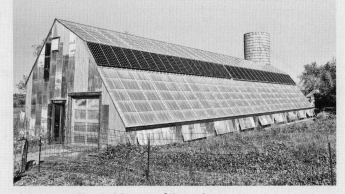

FIGURE 16.5. The Golden Hoof Greenhouse.

porate some water heating for about half the year. They still incorporate the exhaust fans and GAHT system above. The aquaponics system adds a number of components, necessitating a much larger PV system. (For more on aquaponics growing, see Chapter 18.)

- Additional components: 1 air pump (20 W); 1 water pump (40 W); and a water tank heater (400 W)
- Energy consumption per day (aquaponics only): 3.4 kWh (assuming the pumps run 24 hours a day and a water heater runs 5 hours a day on average, throughout the year)

1900s than a farm of today, except that instead of just root vegetables, they are eating home-grown bananas, lemons and figs.

Much of this is possible due to the greenhouse's energy-efficient design and systems. The structure, designed by Alice, an architect, has R-38 insulated walls, and an R-52 roof. Tri-wall polycarbonate forms the south-facing glazing. A Ground-To-Air Heat Transfer (GAHT) system provides much of the climate control, storing heat in the ground. Exhaust fans supplement the many passive vents that line the lower south wall beneath the glazing, and a misting system helps keep things cool in the summer.

The greenhouse also houses one of the farm's power stations. In the center of the south-facing roof, there is a 10 kW solar PV system that powers all of the loads for the greenhouse in addition to some of the outbuildings and the home. Housed in a corner of the greenhouse, the battery bank is comprised of nickel-iron batteries (a less-common choice, which is why we left them out of the discussion above). Though the panels and batteries produce/store enough power to supply the greenhouse's energy, the farm as a whole, is still largely grid-tied. Alice says that if they were to do it again, they would try to go entirely off-grid.

Making a greenhouse completely net-zero-energy and still a tropical forest required a significant investment in structure and systems. The Stareks wanted their slow-food farm to have an extra level of resilience. "We wanted to have fruits and veggies year round, and this is a very brittle environment," says Alice. The Stareks reduced costs where possible by integrating recycled materials and doing much of their own labor. The exterior siding is reused corrugated metal. Old barn doors form the entrances, imbuing the greenhouse with a spunky character that makes it truly one of a kind. More information can be found at the goldenhoof.com.

- Energy consumption per day (total, fans included): 6.2 kWh
- Average energy consumption per year: 2,263 kWh
- Can be supplied by a 1.5 kW PV system

Takeaways

- Passive solar greenhouses (no electricity usage) create quiet and cost-effective backyard greenhouses, but offer less control.
- Many residential growers wire the greenhouse to their home breaker panel, a process easily completed by an electrician or someone with electrical experience.
- If considering adding solar panels to the greenhouse, consider the power demand of your greenhouse and your objectives. Both of these influence the best strategy for the system, whether grid-tied, grid-tied with battery backup, or off-grid.

Further Reading

Marshall, Roger. *How to Build Your Own Greenhouse*. Storey Publishing, 2006. Good information about wiring a greenhouse.

Chiras, Dan. *Solar Electricity Basics: A Green Energy Guide*. New Society Publishers, 2010. Consult this book if you are interested in installing solar panels.

Chiras, Dan. "Off-Grid Aquaponics." DVD. Available from The Aquaponic Source (theaquaponicsource.com) and through the Evergreen Institute (evergreeninstitute.org). All about solar panel systems, especially off-grid systems.

GROWING IN THE GREENHOUSE

CHAPTER 17

Creating the Greenhouse Environment

The soil is the great connector of our lives,
the source and destination of all.
—Wendell Berry, *The Unsettling of America*, 1977

There is a certain excitement that accompanies planning a garden. A border of alyssum will go here; fava beans will trellis up there.… Like a painter envisioning a painting, you can see it in your mind's eye, and there's fulfillment and creativity as you start on this last leg of the journey.

This chapter focuses on the most popular greenhouse gardening method: growing in soil. In recent years, hydroponic, aquaponic and vertical growing methods have become major trends. Due to an impressive wave of recent interest, we've dedicated the next chapter to these methods.

Floor Plans

The floor plan is tied to your planting plan (the layout for what you intend to grow), and your growing method. For residential greenhouses, we strongly recommend raised beds. They offer a much easier gardening experience—they don't require kneeling or bending over. They maximize use of the floor space by creating space-efficient pathways. Most importantly, they allow you to add incredibly rich, nutrient-dense soils to the beds, building up from the native soils in the ground.

In small greenhouses, beds usually line the edges, forming a U with a walkway in the middle. Larger greenhouses can add a center bed, or get more creative, using a "keyhole" bed design, as shown in Fig. 17.1. In any case, a bed layout should make the most efficient use of the floor space, while keeping beds a reasonable width for reaching across. Beds are typically 2'–4' wide and 2'–4' tall. Taller beds allow you to add more high-quality soil for plants' roots (and require less bending over).

Bed height is an important detail to plan early on—when defining your greenhouse geometry and dimensions. We recommend using a south-facing knee-wall below raised beds. Using glazing below the raised beds does nothing for light collection; this area is best insulated as a durable wall. The height of the wall should roughly equal your bed height. Similarly, intake vents—commonly located on the lower south, east and west walls—should be above raised beds, so they are not blocked.

When it comes to greenhouse floors, flagstone, gravel and pavers are good choices for creating walkways between the raised beds. We always recommend leaving the bottom of beds connected to the natural soil. There should be no insulation or flooring underneath beds. Connecting the beds to the earth below ground allows plants' roots to grow as fully as possible (particularly important with perennials), and creates better drain-

FIGURE 17.1.
Raised beds can be made out of a variety of materials (the same options as for an outdoor garden). Concrete blocks are a good alternative to wood. They have the advantage of adding thermal mass to the greenhouse; plus, the openings can form additional planters. Be cautious of pH levels; they can make the soil more alkaline.

age. Since the perimeter of the greenhouse will be insulated, the floor itself does not need to be insulated.

When laying out a floor plan, consider the other ways you might want to use the space. We strongly recommend putting in a work-bench/seeding area, like the one shown in Fig. 17.2. Though many growers don't anticipate it, a bench is very helpful for starting plants, transplanting and propagating. There is also a certain amount of "stuff" that accumulates with a greenhouse. Seed trays, hoses, watering cans and pots can all be stored below a seeding area/workstation.

Integrated Design

Using Animals in the Greenhouse

A warm, protected environment in a greenhouse can be a potential enclave for animals, like chickens, ducks or rabbits. (Integrating fish with aquaculture or aquaponics is discussed in Chapter 18.) But, animals, particularly chickens, can introduce pathogens and unpleasant odors into the greenhouse. For that reason, we don't recommend keeping fowl in the greenhouse.

Some growers see it differently, and advocate keeping chickens or rabbits in an enclosed area in the greenhouse, as they boost CO_2 inside and provide a safe, warm space for the animals. According to Ben Falk of Whole Systems Design, "It's a super safe warm environment with access to fresh greens for baby ducks and geese. They graze right out of the gate even in March, which is huge for their health." You definitely do not want to allow chickens to roam throughout the greenhouse; to them, a greenhouse is a giant buffet (which would leave you without much food).

Rather than integrating animals in a combined space, we recommend using an add-on structure, as in the "greenhouse/chicken coop

FIGURE 17.2.
This greenhouse includes a workstation and sink on the north side.

combination" shown in Fig. 17.3. When joined, both structures are more energy efficient, and it makes excellent use of a single site. The chickens still benefit from the greenhouse, but aren't able to contaminate it.

Worms and Bees

Worms are nature's great fertilizers. Their benefits are numerous and hard to overstate: they accelerate decomposition, aerate the soil, and fertilize it with their castings. Worm castings (a nice name for worm poop) are an excellent source of organic fertilizer for your soil.

To add worms to the greenhouse, we recommend the simple and easy method: adding them directly to raised beds. Some growers go further and build vermicomposting bins in a section of the greenhouse or underneath the walkways. The area directly beneath the pathways of the greenhouse can be turned into a vermicomposting bin, as shown by Jerome Osentowski and other growers. An online search of "vermicomposting pathways" will yield examples and details about how to do it.

FIGURE 17.3.
Chicken Coop and Greenhouse Combination.
Credit: Ceres Greenhouse Solutions

There is also a lot of interest in adding bees to the greenhouse, which serve as natural pollinators. Some growers build beehives into the wall so that bees can freely move outside in the summer and then access the greenhouse in the winter. Like many integrated designs—aquaponics, vermicomposting, chickens—this can be a great benefit to the greenhouse, but is a practice in itself and beyond our scope here.

Without bees, pollinating crops in a greenhouse is typically accomplished by providing gentle air movement. (Wind, just like bees, can carry pollen from the stamen to the pistil of a flower.) Growers can also hand-pollinate by gently shaking flowers.

Hammocks, Hot Tubs, Etc.

In a narrow view, a greenhouse is a place to grow food. In a larger view, it is a place to enhance your health, and that can extend far beyond the food you eat. We encourage residential clients to view the greenhouse holistically—as a space for mental health, relaxation and rejuvenation—in addition to food production. Some clients go as far as incorporating saunas, hot tubs and sleeping areas in the greenhouse. At the very least, we recommend planning some sitting/hanging out space into your floor plan if possible. Once it is up and running, a warm, green and sun-filled space is hard to resist in the winter. Having a place to enjoy the fruits of your labor is a design integration we always recommend.

Building Soils

As the gardener's adage goes, it all starts with the soil. Soil is the foundation of all growth in the greenhouse. Soil determines plants' productivity, and their immunity to pests and disease. It influences the nutrient density and taste of the food you grow. You can build a perfectly controlled environment, but if the soil is not able to grow anything, it will all be for naught. With that in mind, building soils should be a priority when you start growing and something you continue to do year after year.

We say *building* soils, because it usually requires a bit of effort and some amendments to achieve soil that allows plants to thrive. Good soil is biologically active, rich in nutrients, and well textured. We'll break these facets down individually.

Biologically active means the soil has life. Living elements like worms, microbes and mycelium break down organic matter and provide many of the essential nutrients for plants. The "Big 3" nutrients—nitrogen, phosphorus and potassium, or NPK—get a lot of attention. Indeed, these are essential to growth, but so are many other nutrients often called *micronutrients*. Plants need smaller concentrations of micronutrients, but they are still crucial to plant heath. Focusing solely on NPK is a limited view that can hinder the success of your greenhouse.

The best way to give plants the full spectrum of nutrients they need is by creating biologically active soils rich in organic matter (sources of carbon and nitrogen, among other things, contained in plant matter). Organisms in the soil make nutrients available to plants by breaking down organic matter. They also help aerate soil. In their wake they leave behind pockets for air and water and for plant roots to penetrate further into the soil. Plants have evolved to depend on soil organisms, and there is no easy replacement for their complex biological role. Nutrients can be supplemented with liquid fertilizers or soil amendments, though these are not a direct replacement for biologically active soil.

The Importance of Organic Matter

How do you encourage biologic activity? The easiest way is to add lots of organic matter to soils, following the mantra "if you build it, they will come." You can purchase compost (decomposed organic matter and sometimes manures) from a reputable source, or simply make your own. Our society discards immense quantities of organic material. Bags of leaves on the curb, nitrogen-rich coffee grounds from your local coffee shop…there are numerous sources for organic materials. Disrupting these waste cycles and turning them into rich, organic soil is a positive and satisfying pursuit.

Creating your own compost simply requires organic material, water, air and time. You can fine-tune your organic recipe by researching particular organic materials and finding out whether they are good sources of carbon or nitrogen. See the books listed in Further Reading for more on composting.

The main drawbacks, in our view, are simply the time and labor involved in creating large amounts of compost. Raw organic materials usually take many months to decompose before they are good soil amendments. We recommend starting a backyard composting pile well before you build your greenhouse; that way, you'll have a source of compost by the time the raised beds are installed. Some growers, like Jerome Osentowski of the Central Rocky Mountain Permaculture Institute, add raw organic matter directly to the growing beds, building soils right in the greenhouse. It still takes time for the material to decompose in the beds, but there is less work involved in moving materials. Jerome's book, *The Forest Garden Greenhouse*, provides more information on building soils.

Though more labor intensive, DIY composting saves money and usually creates a better product at the end. Compost sitting out at your site will naturally be inoculated with microorganisms and mycelium, and these beneficial soil organisms will then be transplanted to your raised beds.

While composting and organisms add nutrients to the soil, it's impossible to know the exact nutrient makeup of a soil just by looking at it. To do that, we recommend getting a soil test (provided by many states' agricultural extension offices). These will tell you the soil's pH, nutrient density and structure (percentages of sand, silt and clay). Most plants like a neutral pH, around 7.5.

After you have your soil test results, you can respond with soil amendments to supplement for a deficiency. There are a huge range of options: kelp, fish emulsions, blood meal, bone meal and many others can bring out specific qualities in the soil and thus, plants. This takes some additional research; we recommend looking into options once you know the profile of your soil.

While we recommend doing as much composting as is logistically feasible, it is a tall order to fill the entire volume of growing beds with your own compost (limitations are usually space and back power).

Thus, many growers wisely purchase a base soil substrate (often a mix from a landscaping company) and use homemade or purchased compost to amend.

Purchasing Soils

Landscape companies typically offer something like a "top soil mix" or a "compost mix." While you can use these to fill some of the volume of the beds, be cautious about overusing them. Many are very high in salts from manures. Ask for a copy of the provider's soil test to know what you are starting out with. You may be able to source local compost (from a farm, for example). Generally, asking other gardeners in the area is the best way to find the best local sources.

A second common source of material is the bagged soils you find at garden and hardware stores. These mixes can be useful for amending soil—particularly to enhance drainage—but we don't recommend relying on them extensively. First, they are extremely expensive per volume. They are also pretty wasteful, given the long-distance shipping, and not always sustainably sourced. Finally, though they say "soil," these mixes are usually predominantly peat moss or coco coir, fibrous materials that make soil light and fluffy. (Manufacturers do this because they're cheaper to ship, and most of these mixes are marketed as well-draining potting soils.) That makes them low in organic matter, the basic building block of nutrient-rich soils.

What about using the existing soil underground? Perhaps you have some extra soil left over from when you installed a GAHT system. Using your native soils is a viable option; it all depends on what kind of soil you are starting with. Here in Colorado, soils are very heavy in clay, which prevents them from draining well. So, if we want to use native soil in a raised bed, we have to add substantial amounts of organic material.

This brings us to soil texture. The percentage of sand, silt and clay in a soil affects how it holds water and air. It changes the feel of soil and how easily roots can penetrate it. Unlike nutrient compositions, you can get a feel for soil texture just by touching it. If you rub a clump of soil between your fingers and it feels slimy and smooth, it probably has a high concentration of clay. High clay content restricts drainage, which can lead to soils becoming waterlogged. This in turn cuts off air to plant roots. In a humid greenhouse environment, well-draining soils are critical. These can be created by adding organic matter or an amendment like perlite or coco coir. A soil test can tell you the structure of your soils to know whether this is needed.

If this all seems overwhelming, focus on these few takeaways when building good soils: First, you can't really overdo it with organic matter, as long as it's properly composted and not too high in salts (from uncomposted manures). Secondly, keep soils well-drained. Finally, use a soil test to evaluate the nutrient concentration and decide if you need further amendments to target a specific deficiency.

Qualities of Good Soil:

• Rich in organic matter	• Rich in nutrients
• Healthy biologic activity	• Low in salts
• Drains well	• Neutral pH

Planting Plans

Each planting plan is unique, though there are some considerations that apply to all growers. Solar greenhouses have *microclimates*, zones in which light and temperatures vary. These change throughout the year as the angle of the sun changes. Figure 17.4 show typical microclimates during the winter. Anything located near the glazing will experience the greatest temperature swings, and coldest temperatures. At night, plants here are inundated with cold air. As air cools around the glazing and descends, it eventually collects at plant level. For that reason, we recommend planting cold-tolerant crops along the front beds in the winter.

The back of the greenhouse, particularly if there is thermal mass, will be the warmest and most stable in temperature. It is a good area for more sensitive plants, and those that can take advantage of the vertical space. Tomatoes, cucumbers, and peas are often placed here, and allowed to trellis up the wall.

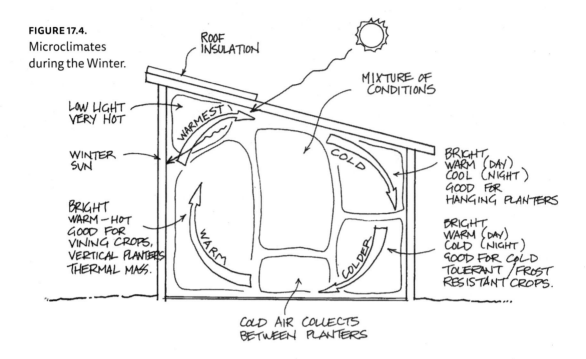

FIGURE 17.4.
Microclimates during the Winter.

ROOF INSULATION

MIXTURE OF CONDITIONS

LOW LIGHT VERY HOT

WINTER SUN

BRIGHT WARM – HOT GOOD FOR VINING CROPS, VERTICAL PLANTERS, THERMAL MASS.

WARMEST

COLD

WARM

COLDER

BRIGHT, WARM (DAY) COOL (NIGHT) GOOD FOR HANGING PLANTERS

BRIGHT, WARM (DAY) COLD (NIGHT) GOOD FOR COLD TOLERANT / FROST RESISTANT CROPS.

COLD AIR COLLECTS BETWEEN PLANTERS

The temperature differences between these zones can be quite noticeable in the winter. In the summer, conditions are more uniform, especially when there is some shading on the north wall (from roof insulation or shade cloth).

Considerations for Creating a Planting Plan

- **Insectary plants:** Insectary plants attract beneficial insects, which in turn, naturally fend off harmful insects. Attracting and keeping these insects is the basis for *integrated pest management,* or IPM, a holistic approach to pest control. Insectary plants are often flowers and/or herbs, which have the additional benefit of adding color and aroma to your garden. Adding them to your greenhouse helps create a resilient ecosystem that mimics nature. They sustain beneficial insects throughout the winter (when there is no access to outdoor food sources). In the summer, they invite pollinators and beneficial insects into the greenhouse to help with pest management.

Plants for Beneficial Insects

- Alyssum
- Composite flowers (daisy and chamomile)
- Chives
- Dill
- Fennel
- Marigold
- Mint

- **Nitrogen-fixers:** Just like in the outdoor garden, certain plants can help add nitrogen to the soil. Beans, peas and clovers are among the plants that absorb nitrogen from the atmosphere and deposit it into the soil through their roots. They can be interspersed throughout the greenhouse to enhance nutrient rich soils.

- **Plant height:** A bit obvious, but sometimes forgotten factor: plants grow to different heights and can shade each other. During the winter, we generally recommend placing tall, vining plants on the north wall, and shorter plants in the front. In the summer, tall plants can be used to help control heat gain; you may want to use their shading effect as they trellis up the south wall. Or, you can intersperse tall and short, using shorter shade-tolerant plants as ground cover, which is more characteristic of natural outdoor environments.

- **Mimic nature:** Natural environments are diverse polycultures. Biodiversity helps thwart the spread of pests and disease by creating resilient ecosystems. For example, if you have a single large area of lettuce starts, and aphids establish there, they are very susceptible to the pest outbreak. If the lettuce is interspersed with other plants—perhaps some mature plants more resistant to pests, or flower that attract beneficial insects—the system as a whole is more resilient. Diversity is nature's defense system and should be designed into the greenhouse as well. (Commercial greenhouses are sometimes the exception, though even those can take advantage of diversity and permaculture principles.)

- **Think three-dimensionally:** In his book, *The Forest Garden Greenhouse*, Jerome Osentowski said it best: "You're sculpting a three-dimensional landscape." Consider not just how plants will occupy the area of the greenhouse, but the entire volume. Do you want to create multiple stories of plants, or stack tall plants on the north side? Jerome has become a master of creating three-dimensional landscapes in his greenhouses, interspersing fruit trees, vining crops, trellising crops (like passion fruit) and annual vegetables.

- **Experiment:** The year-round environment of an efficient greenhouse opens up the possibilities of what you can grow. It's an opportunity to branch out and try new crops. Ever eaten a tree tomato or grown purple lablab beans? Now you can. Seed catalogs, traveling to other areas, and talking to other growers are great sources of inspiration.

Planting Schedules

Year-round greenhouses are constantly evolving environments. Growing is a continual process of life and death, seeding and harvesting. As a backdrop to the daily tasks, life in the greenhouse also takes on seasonal rhythms. Each season has certain characteristics and corresponding tasks depending on the light and temperature afforded by the sun. A simplified summary is shown in Fig. 17.5.

The winter is predominantly characterized by low light and very slow growth. We call this "hibernation mode." Plants survive, but their growth slows to a snail's pace when the daily light integral drops to the low single digits. (See Chapter 5 for more on light requirements.) Low light is especially detrimental to seedlings. To accommodate, start plants early enough (the late summer/early fall) that they can persist and still provide a good harvest through the winter. Plants should have full leaves and developed root systems before day length declines drastically, around October/November in North America. That way, they can be harvested through the winter, providing fresh produce that is impossible to find at the grocery store.

Of course, the exact timing varies by crop and your latitude. A good indicator of when to start planting for winter is when there are still 11 or 12 hours of sunlight (not an exact number, but a rough approximation for solid growth). For us, in Denver, at 40-degree latitude, that corre-

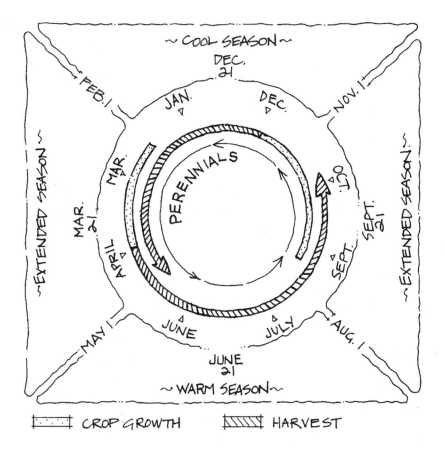

FIGURE 17.5. Simplified Year-round Growing Cycle. This is a generalized depiction intended to show the "second" season afforded by year-round greenhouses; that which occurs while outdoor gardens are fallow. The specifics of your planting calendar are entirely dependent on your goals, greenhouse and climate. Credit: Adapted from "A Solar Greenhouse for the Northwest," by Ecotope, Inc.

sponds to September to mid-October in the fall (and mid-February to March in the spring).

A mistake we've made a few times in our own greenhouse gardens is to wait too long before starting plants for the off-season. When you are up to your elbows in tomatoes in August and September, the last thing you think about is starting more plants. But if you miss this window, it is hard to recover without adding supplemental lighting or purchasing starts. Thus, keep in mind the planting calendar in Fig. 17.5 and plan ahead.

We recommend starting cold-tolerant crops for winter growing, mainly because these are usually also shade-tolerant or low-light crops. If your greenhouse runs warmer, you can keep warm-season crops like tomatoes, cucumbers and peppers going, though they don't produce much fruit in low light. However, the occasional vine-ripened juicy red tomato in the dead of winter may be well worth it. Because it is hard to tell what will thrive in your winter conditions, we recommend starting a range of plants and seeing what does well the first year.

In spring, any remaining annuals that were started in the fall experience robust growth as light returns. The greenhouse becomes a verdant oasis at a time when most other growers are still hankering to get outside to start their gardens. (We sometimes call spring the "jealousy season.") It's an excellent time to start plants for growing on in the greenhouse or transplanting outside.

That brings us to summer. Some growers grow heat-tolerant crops in the greenhouse in the summer. Others let it go fallow, concentrating on their outdoor gardens instead. Either way, things typically get shifted around during the summer as some plants get moved outside.

As a backdrop to these cycles, perennials (fruit trees, flowers and herbs) can remain in greenhouse year-round, adding to the abundance. You may find that a greenhouse allows many annuals to turn into perennials. In our demonstration greenhouse at Ceres, a kale plant seemed to like the environment, so we just left it. After a couple of years we had a "kale tree" about four feet tall with a stalk several inches in diameter. Though the leaves got tough in its old age, it kept producing, providing a nutritious supplement to green smoothies.

Crop Scheduling

In his book, *The Winter Harvest Handbook*, Eliot Coleman astutely makes the point that the "cross-quarter" days are very useful for crop planning. Most of us are familiar with the "quarter days"—the solstices and equinoxes. However, the cross-quarter days—the mid-point between the equinox and the next solstice—align more closely with the beginning of each season and can serve as a rough guide for your planting schedule:

- Feb. 1: Start warm-season annuals as days start to lengthen.
- May 1: Begin to transplant starts outdoors, depending on your last frost date and the crop.
- Aug. 1: Start planning for cool-season annuals for the winter growing season, though they may be started a little later depending on your climate.
- Nov. 1: Plants for winter harvesting should be nearing maturity by this time.

Obviously these dates are not concrete, and vary by climate. They're meant to serve as reminders for the next phase of growing...a gentle ping that the season is changing and it's time to prepare.

FIGURE 17.6.
Super-size Collard Greens. Credit: Flourish Farms

Observation and Documentation

Year-round growing isn't a requisite for a solar greenhouse. Three-season greenhouses allow growth most of the year, but go dormant in the winter or the summer. Other growers choose a middle road, keeping some frost-tolerant crops in the greenhouse in the winter without actively trying to maintain certain temperatures. Fig trees, for example, can withstand temperatures down to about 10°F (−12°C) and can be left in a greenhouse, mostly dormant, for the winter with little tending.

Your growing calendar and planting plan will be completely unique to you. We recommend using the experience of others and experimenting on your own to see what works best for your greenhouse. Documentation is extremely helpful in this regard. It hastens the learning process, allowing you to get a full picture of how the greenhouse operates throughout the year. Simple journal entries can be extremely helpful to record observations. For example:

November 15, 9 AM: Greenhouse 50°F. Cold damage on basil. Re-seeded spinach.

May 5, 12 PM: Greenhouse 90°F. Tomato starts taking over. Start these later next year.

Simple notes can go a long way to clarify patterns. Other growers prefer a more organized chart for record keeping. Recording climate data like temperature, light and humidity, with thermostats and data loggers can also be extremely helpful. Whatever your style, documentation is a good habit to get into. It can inspire you to observe more and see patterns that emerge. That, in turn, allows you to anticipate changes and refine your planting schedule year after year.

Pest Management

Pest management sections of books tend to focus on introducing beneficial insects or using pesticides after a pest problem emerges. These are emergency measures, very useful, but only part of the full picture. To us, pest management starts long before harmful insects get into the greenhouse. It starts with creating a resilient ecosystem that naturally resists pest outbreaks on its own. This involves using polycultures to create diversity and adding beneficial plants to your planting plan. It equates to focusing on the full picture of plants' health. When plants are stressed, their natural defense systems are compromised. By building

---

 Should I use screens over windows/fans?

We don't advocate using screens in the greenhouse. Most harmful pest populations, like aphids, can find their way in with or without a screen. Moreover, screens tend to block out many beneficial insects, like ladybugs, which are larger and can't get through.

Commercial greenhouses that use monocultures are a different story and may need to side toward the "sterile-environment" approach, using screens and other measures to prevent all insects from entering the greenhouse.

---

good soils, maintaining proper watering, and preventing overheating and cold shock, you enable plants to thrive. In turn, they are able to naturally deter pests on their own (to an extent). Hence, pest management begins with the greenhouse itself, including design and climate control.

This approach is more akin to a philosophy than a management strategy. Some growers try to create a sterile, extremely controlled environment in the greenhouse. Indeed, it is wise to take some measures to prevent pests from entering the greenhouse (like washing off starts purchased from garden centers). We find that insects—both good and bad—are natural and inevitable parts of the garden. For most residential greenhouses, management doesn't mean complete annihilation of pests. Rather, it is about controlling their populations using natural defenses and healthy plants.

Unlike an outdoor garden, pests in a greenhouse find a haven free from many natural predators. Their populations can easily explode if not caught early. Observation is critical. If a population is caught early, it can be quickly remedied with simple, natural methods. Removing dead plant matter/debris and checking plants (particularly under leaves) will go a long way to preventing problems. Observation also includes correctly identifying what the problem is. The top three pests that afflict greenhouses are aphids, whiteflies and thrips, and each requires a tailored control strategy.

This brings us to a third level of management in the case of a pest outbreak. There are a range of natural techniques, such as using beneficial predatory insects or organic sprays. Ladybugs or lacewings will prey on aphids, naturally controlling their populations. Ladybugs can consume 5,000 aphids in their lifetime, and quickly reproduce, creating alligator-looking larvae, which in turn consume more aphids. Some growers use insectary flowers to try to maintain a year-round population of ladybugs in the greenhouse. Others purchase ladybugs (online or at garden centers) if an aphid outbreak occurs. Lacewings are also a popular choice, considered by some to be more effective than ladybugs.

Organic sprays are also useful, if needed. Pyrethrum is a naturally occurring oil that can be used as a pesticide. An interim (and much cheaper) step is simply knocking pests off plants with water. Gently spraying affected leaves can disrupt resident aphid populations enough to curtail their growth.

Water

Getting Water to the Greenhouse

The water requirements of greenhouses vary greatly throughout the year. For most of the year, greenhouses require much less water than outdoor gardens, where winds dry out the soil. The protected environment of a greenhouse traps moisture—often, too much moisture—making the watering requirement much lower for most of the year. The exception is the summer. If the greenhouse is operating in the summer, it will need to be constantly ventilated to avoid overheating. As a result, regular watering is a necessity.

Getting water to the greenhouse can happen in three ways: collecting rainwater, bringing in a hose from the outside, or installing a conventional water line. Collecting rainwater from the greenhouse roof is an excellent method that enhances the self-sufficiency of the structure.

Rainwater collection can be done with gutters and drains, similar to how it works with a home. There are two main options. With exterior gutters, water flows from the gutters through a drain through the wall and is collected in a reservoir inside, as shown in Fig. 17.7. Alternatively, the gutters can be inset into the roof and drain down to barrels inside the greenhouse (these are usually located on the south side, since water runs off the angled south-facing roof).

If a small residential greenhouse is located near the home, a hose can be brought out to provide watering as needed. While very simple and free to set up, this can get cumbersome as time goes on, particularly in the winter, when hoses should be disconnected from the residential plumbing in freezing climates. To address this issue, one of our clients devised an elegant solution for storing domestic water in the greenhouse. Shown in Fig. 17.2, a large water storage tank is located on the wall inside. This connects to a valve that can be connected to a hose outside. When the hose is on, the water pressure is sufficient to fill the

FIGURE 17.7.
Designed by Dirt Craft Natural Building, this cob greenhouse uses rainwater catchment.
Credit: Dirt Craft Natural Building

water tank. Once the tank is full, water can be gravity-fed from the tank to watering cans...easy access without taking the hose out every day.

The final method of getting water to the greenhouse is a conventional water line. In this case, a copper or PVC pipe is buried underground, bringing water from a nearby structure to a spigot inside or outside the greenhouse. The water line should be buried below the frost line in locations with freezing winters. The line must be plumbed to the house. Because most municipal water sources contain chlorine—harmful to plants in high concentrations—we recommend adding a water filter to the end of the line. Often, a plumber completes this process, but there are DIY resources available to help you install a line yourself (see *How to Build Your Own Greenhouse* by Roger Marshall).

Watering Systems

Hand-watering and drip irrigation systems are the conventional options. Drip irrigation has the advantage of providing regularly scheduled watering, helpful if you plan to leave the greenhouse for a few days on its own. The setup is the same for outdoor systems: a main hose, fed by a water pump, brings water to a series of emitters that are spread within the beds. It takes a bit of monitoring to ensure that water is evenly distributed and getting to the plants in need.

Wicking beds are another watering option that have a number of advantages. These are raised bed containers with a water reservoir on the bottom. Cloth wicks extend from the reservoir and disperse water

FIGURE 17.8.
Greenhouse with wicking raised beds from Farm Tub™. You can see the cloth wicking material in the unfilled grow bed.

into the soil. This reduces water usage by watering plants' roots directly (not the surface of the soil). Thus, it reduces evaporation from the soil. That, in turn, reduces humidity in the greenhouse and the watering requirement. Self-wicking beds can be made from basic containers and materials or purchased pre-made, like the Farm Tub™ wicking beds shown in Fig. 17.2 and 17.8.

Further Reading

Lowenfels, Jeff and Wayne Lewis. *Teaming with Microbes: The Organic Gardener's Guide to the Soil Food Web*. Timber Press, 2010.

Lowenfels, Jeff. *Teaming with Nutrients: The Organic Gardener's Guide to Optimizing Plant Nutrition*. Timber Press, 2013.

Osentowski, Jerome. *The Forest Garden Greenhouse*. Chelsea Green Publishing, 2015.

Smith, Shane. *The Greenhouse Gardener's Companion: Growing Food and Flowers in Your Greenhouse or Sunspace*. Fulcrum Publishing, 2000.

CHAPTER 18

Aquaponics and Hydroponics

Hydroponics, aquaponics, aeroponics, vertical growing.... Today, there is a huge range of innovative growing methods that can be incorporated into a solar greenhouse. In fact, most are particularly well suited for greenhouses because they involve equipment and electronics that need protection from the elements. Solar greenhouses provide a controlled, year-round environment that increases yields and maximizes investments.

This chapter explores the ways to grow beyond soil gardening. We'll focus on a trend that's becoming increasingly popular: aquaponics, the practice of raising of fish and plants in a symbiotic ecosystem. We'll start with the original "ponics" method: hydroponics.

Hydroponics

From the Latin roots for *hydro* (water) and *ponos* (labor), hydroponics literally means *water working*. Plants' roots are suspended in a nutrient solution instead of soil. By adjusting the solution, the grower can precisely regulate the nutrients, water and air the plant receives. This gives growers a level of control over the plant environment unattainable with soil gardening, which involves complex interactions among biological elements and the soil.

Another advantage of hydroponics is that it allows plants to be grown more densely since they don't need the large volumes of soil from which to acquire water and nutrients. Plants are either grown in a small volume of an aerated nutrient solution or in a growing medium like a porous rock or expanded clay. This makes the growing medium much lighter than soil growing.

Without the large weight of soil, gardens can be more easily integrated into other infrastructure. Hydroponics opens the possibility for rooftop greenhouses and vertical growing, enabling farms to crop up where they would otherwise not be possible (such as city centers, office buildings, indoor classrooms, etc.).

Finally, like all the "ponics" systems in this chapter, hydroponics is commonly touted as having greater overall yields. Tighter planting density, vertical planting, and complete control of nutrients make for greater overall crop production than traditional soil methods. Tawnya Sawyer of Colorado Aquaponics notes that greater plant size is partly due to the energy of the plant not being wasted on the hard work of pushing roots through heavy soils. Instead of searching out nutrients in possibly depleted soil, or waiting for water (or receiving too much water), all the plant's energy is put into growing edible leaves, flowers and fruits.

The advantages of hydroponics—more control, higher yields and clean environments—have made it a mainstay in our commercial agriculture system. According to the USDA, nearly all tomatoes grown in large commercial greenhouses the US and Canada are grown hydroponically.[1]

Though popular, hydroponic growing comes with its drawbacks. It lends itself to growing a single crop in large-scale monocultures because all the plants in one bed/container are treated with the same water and given the same inputs. Large-scale monocultures increase the risk of pest problems.

Many hydroponic nutrient solutions are manufactured from mined minerals or synthetic chemicals. Organic hydroponic nutrients are made from bat or seabird guano, seaweed extracts, fermented vegetable waste and other biological or mineral elements. While these products can be sustainable and organically certified, they create other challenges such as pH management, sludge buildup, and plumbing clogs.

Finally, hydroponics requires research, learning and a larger budget, usually. The growing infrastructure (including plumbing systems, growing beds and media) increase the cost. It also takes time to learn. Given those pros and cons, we find it suitable for commercial growers looking for its essential advantages—more control, higher yields and integration into existing infrastructure. It can also take advantage of unused height in the greenhouse, using vertical systems like the WindowFarms mentioned below.

Hydroponic Growing Systems

Most hydroponic growing occurs in the commercial sector, with large-scale monocultures being grown in huge greenhouses, but the trend is making inroads into residential and small-scale commercial spaces. Henry Gordon Smith of Blue Planet Consulting, a Brooklyn-based consulting firm, advises residential growers to build their own system. "It's not only cheaper but allows you to understand the mechanics of the system and correct problems if things happen down the road."

There are a few hydroponic kit systems as well, such as the ones provided by WindowFarms. Based in New York, WindowFarms' mission is to enable city-dwellers to turn their windowsills into gardens using hydroponic growing towers. They offer some complete hydroponic growing systems. Mostly, they act as a forum for growers to exchange ideas on building their own hobby systems using recycled materials like soda bottles.

Aquaponic Gardening

Aquaponics takes hydroponics and adds fish to the equation. Aquaponic systems grow fish and plants together in a symbiotic ecosystem. In the basic aquaponics cycle, the wastewater of the fish tank (containing natural fertilizer) is fed to the plants. The plants filter the water, removing toxic nitrates, before it's returned back to the fish, creating a closed loop. A third actor is a living bio-filter made up of worms and bacteria in growing beds. These serve as a keystone element: they convert the fish waste into a useable fertilizer for the plants, allowing the two systems to grow, symbiotically, as one.

By using fish waste as fertilizer, aquaponics eliminates the need for petroleum-based fertilizers. It eliminates the excessive fish waste pollution that arises from conventional aquaculture. Similar to hydroponics, aquaponics systems use only one tenth the water compared to soil-based gardens, since water is continually recirculated. And, at the end of the day, you not only get fresh produce out of the greenhouse, but a

Vertical Growing Systems

Growing up, not just out, allows you to maximize use of the greenhouse space and dramatically increase production. Consider that a 12' × 20' greenhouse has 144 sq. ft. of floor space—but 2,160 cubic feet of *volume* (assuming an average wall height of 9'). It's not practical to fill every cubic foot with plants, but it is easy to take advantage of under-utilized growing spaces in the height of the greenhouse.

FIGURE 18.1. Soda Bottle Towers.

FIGURE 18.2. ZipGrow Towers.

safe chemical-free source of protein. For those people who want to be more self-reliant in growing their own food, fish are far less noisy and odorous than most animals.

So what's the catch? While aquaponics presents huge benefits for our health and the environment, it is a more complicated growing method. In soil-based gardening, there is one set of relationships occurring

---

A prime area for vertical growing is the north wall. It receives full light for most of the year, and tall plants there won't shade other plants; it also provides a structure for trellises and hanging planters. Some plants naturally take advantage of vertical areas, such as trees and vining crops. Vertical systems can assist as well, allowing you to grow small, high-value crops like herbs in the higher reaches of the greenhouse. Vertical planters can be soil-based or hydroponic (or integrated with an aquaponic system) and range from very simple to extremely elaborate. A few examples:

- **Recycled soda bottle planters:** If you want to get lost for an afternoon, do a search for "vertical growing" on a social media site like Pinterest. A plethora of creative ideas will emerge, like the recycled soda bottles planters pictured in Fig. 18.1. Each bottle is cut and filled with soil, or, if growing hydroponically, a non-soil substrate like expanded clay. A pump feeds water to the top, and holes in each bottle allow water to percolate down. This is a great way to make use of the unused corners of a greenhouse, like the basil growing in the southeast corner of a residential greenhouse.
- **Aeroponic towers:** Tower Garden™ is a product that has popularized aeroponic growing. Plant roots are suspended in air and regularly misted by a nutrient solution, rather than a stream of water, as with hydroponics. Tower Gardens are available for both home and commercial growers.
- **Hydroponic towers:** Hydroponic growing columns are also possible for commercial systems. Wyoming-based company Bright Agrotech builds ZipGrow Towers, in which plants are grown hydroponically between recycled plastic mats inside a plastic post, shown in Fig. 18.2 Their hope is that with these modular vertical growing systems, anyone can start a small farming operation with just a small space.

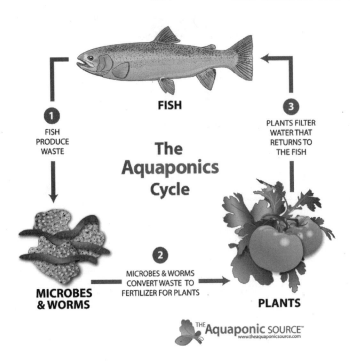

among plants, soil, microbes, insects and the environment. Aquaponics introduces two more ecosystems with their own needs: fish and bacteria. Like plants, fish have certain requirements—temperature, pH, nutrients and oxygen levels. Keeping these factors consistently balanced takes monitoring and daily attention. If fish suffer, the plants in turn suffer, and the whole system can break down.

This is not to say you shouldn't consider aquaponics, only that it's an endeavor in itself, on top of managing a greenhouse. Many growers wisely take a one- or two-day course

FIGURE 18.3.
The Aquaponics Cycle.
Credit: The Aquaponic
Source

before diving in. Research is necessary, and lots of tinkering and adjusting should be expected.

For more on the basics of aquaponics, see a book like *Aquaponic Gardening* by Sylvia Bernstein. Our purpose here is not to explain aquaponics systems themselves, but rather to elucidate how to successfully integrate aquaponics into a greenhouse. We are looking at aquaponics through the lens of greenhouse design, focusing on making the system as energy efficient as possible.

Aquaponic Greenhouse Design

There is a natural synergy between aquaponic growing and solar greenhouses. Aquaponics incorporates fish tanks that can double as thermal mass for the greenhouse. Likewise, a solar greenhouse creates a thermally stable environment conducive for growing fish and bacteria, which need stable temperatures in a narrow range. Aquaponics requires a controlled environment, and an energy-efficient greenhouse is well suited.

Tank Placement and Floor Plan

Like any source of thermal mass, fish tanks are most logically placed on the north wall of the greenhouse where they can absorb direct light during the winter. In very large greenhouses, they can be housed in a separate, insulated room that can be kept at a more stable temperature.

Fish tanks come in all shapes and sizes—rectangular IBC totes, circular stock tanks, even bathtubs. Whatever your choice, sketch the tanks into your floor plan at the start of the design process. Large circular tanks can be an inefficient use of space in a rectangular greenhouse, but some growers prefer them. There is much discussion in the aquaculture industry about the pros and cons of circular versus square, rectangular, or raceway-style fish tanks. Because there aren't any absolutes on tank shape, you'll need to make your own determinations based on layout, space, tank size and what's available.

Ensure that your floor plan is big enough for both growing beds and fish tanks. If using a media bed system, the standard rule of thumb is to have a 1:1 ratio between the volume of the growing beds and fish tanks according to Sylvia Bernstein, author of *Aquaponic Gardening*. In other words, for every cubic foot of grow bed there should be one cubic foot of water in the fish tank. As fish mature and you get used to your system, you can add additional grow beds, expanding to a 2:1 ratio.

Following this advice, your fish tanks occupy the same amount of space as your growing beds. For residential greenhouses, that makes space very tight inside. Think carefully about how to maximize floor space so you can still grow as much as you'd like, or be aware you may need a larger greenhouse to accommodate tanks.

To help save space and stabilize the temperatures of the tanks, many growers choose to bury tanks partially underground. This also helps insulate the tanks, as the soil stays a more stable temperature than the air. This decision is based on your aquaponics setup. In some systems, like a "basic flood and drain," water is gravity-fed from the growing beds to the fish tanks, which allows for easy placement of the tanks underground, beneath the beds. (See color section, Growing Power, for an example.)

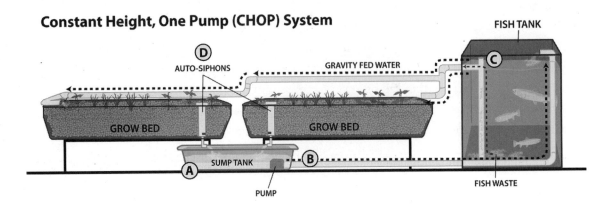

Constant Height, One Pump (CHOP) System

FIGURE 18.4.
Constant Height
One Pump (CHOP)
Aquaponics Growing
System. Credit: Sylvia
Bernstein

Other system designs place the fish tanks *higher* than the grow beds. Water drains via gravity into a sump tank before it's returned to the fish tank. In this design, called "Constant Height, One Pump," or CHOP, the sump tanks sit below the grow beds, as shown in Fig. 18.4. This makes it easy to partially sink sump tanks underground.

If considering burying tanks, you must leave (or create) the necessary openings in the floor. Flooring in an aquaponic greenhouse is normally a concrete slab. Concrete gives a stable, level base for fish tanks, which are incredibly heavy. Gravel, flagstone or pavers are other options. Keep in mind that aquaponics involves plumbing, which can be laid on top of the floor, but this creates a minor tripping hazard. This is an advantage of using something like pavers; plumbing can be installed under the floor but still be accessed, if necessary, by lifting up the pavers.

Insulate Fish Tanks

In climates with cold winters, fish tank heating is typically the #1 energy expense of an aquaponics greenhouse. Fish can't regulate their own body temperature; they depend entirely on water temperature. The common rule of thumb is that the fish tank temperature should not fluctuate more than three degrees in a 24-hour period. If their body temperature fluctuates dramatically in a short time period, fish become stressed and can die.

Fortunately, water is a forgiving material in terms of maintaining stable temperatures; its high heat capacity means it resists quick fluctuations. Still, the narrow margin for error means most fish tanks require heating in cold climates. Fish tanks can be insulated by wrapping a malleable insulating material (like bubble wrap) around them. If growing a warm-water species, tanks should be insulated even if they are buried underground. Remember that soil temperatures deep underground average about 50°F (10°C) across North America, and your fish may require warmer water than this. If above ground, tanks should be insulated underneath as well, not sitting on a floor that is colder than the tank.

In addition to the tanks themselves, we recommend insulating the plumbing and everything that can lose the system's heat. Pipe insulation, as shown in Fig. 18.5, is very low cost. Simple steps like these are the low-hanging fruit when it comes to growing in cold climates with aquaponics. According to Jeremiah Robinson, longtime aquaponics grower in Madison, Wisconsin, "when growing in cold weather, managing your heat loss means the difference between spending $1,000 on winter heating and spending $100."

FIGURE 18.5.
Plumbing Insulation.
Credit: Frosty Fish
Aquaponics

Jeremiah goes even further to reduce tank heating costs as much as possible. He converts old freezers into fish tanks to create super-insulated, and almost airtight, fish tanks. According to Jeremiah, sealing fish tanks greatly helps reduce heat loss from the tanks during the day. The reason goes back to phase changes, discussed in Chapter 12: During the day, water evaporates off the surface of open tanks as the greenhouse warms up. This has a major cooling effect due to the energy absorbed by the phase change of water to vapor. Open fish tanks act like small evaporative coolers during the day.

Sealing fish tanks requires modifications so they can be sufficiently aerated. For more on Jeremiah's freezer fish tanks see frostyfish.com.

In other climates, the cooling effect during the day may not be problematic and may even be a benefit. If the greenhouse is well ventilated, evaporating water can be a cooling mechanism for hot-climate greenhouses, though fish tanks will need to be topped off with water more often.

However you choose to house your tanks, we recommend monitoring humidity levels. Particularly in the winter when the greenhouse is closed, tanks can add a great deal of water vapor to the air, which has negative effects in excess. Putting a loose covering of polycarbonate over a tank is a good tactic for cutting down on evaporation during the day.

Select Fish for Your Climate

The fish species you select impacts how you run your greenhouse, and thus your final design. We recommend choosing a fish species that thrives in conditions your greenhouse will naturally gravitate toward. Trout for instance, are cold water fish that can tolerate water temperatures down to 35°F (2°C), and are much easier to grow in colder climates. However, they can suffer and die quickly if greenhouse temperatures exceed 70°F (21°C), so they are a poor choice in the summer months in many locations.

While these recommendations seem simple, the challenge is that temperatures in most greenhouses fluctuate seasonally and span more

FIGURE 18.6: Temperature Ranges of Common Species of Aquaponics Fish.

Species	Warm or Cool Water Preferred	Temperature Range Tolerated	Temperature Range for Optimal Growing
Tilapia	Warm	60–95F (16–35C)	74–80F (23–27C)
Catfish	Warm	32–95F (0–35C)	75–85F (24–29C)
Yellow Perch	Warm	63–79F (17–26C)	70–75F (21–24C)
Hybrid Striped Bass	Cool	45–75F (7–24C)	60–72F (16–22C)
Trout	Cool	35–68F (2–20C)	55–65F (13–18C)

than one species' range. In these cases—if multiple fish species are possible—we recommend siding with the warmer-water fish for a number of reasons. It's much easier to heat water than to cool it. There are myriad ways to heat water, including using solar hot water, but few ways to remove heat from a fish tank. (Chillers are expensive equipment.) Secondly, bacteria and worms prefer warmer water (between 63°F–93°F [17°C–34°C]). Thirdly, there are more edible crops that like (or at least tolerate) warm water over cool water; they will grow faster in those conditions. Remember that the fish tank water (whether warm or cool) will be used to water plants. In short, if your greenhouse overlaps with multiple fish species' requirements, then side with the warm-water fish.

Of course, other factors come into play when choosing a fish species: their growth rates, what you like to eat, what is in demand if growing commercially, what is easily sourced in your location, etc. These may trump climate considerations, in which case design your greenhouse to best accommodate the fish you select. For instance, if you choose a warm-water species and live in Canada, invest heavily in insulation and efficiency for the greenhouse. A sustainable water heating method like solar hot water to provide water and space heating would be very beneficial (provided enough sun).

Controlling Air and Water Temperature

It is also important to regulate air temperatures to sustain an active, healthy bacteria population in the growing beds. Bacteria thrive in warm conditions, between 63°F–93°F (17°C–34°C), their reproduction slows dramatically in colder conditions. They die in freezing conditions or above 120°F (49°C).

Thus, regardless of fish species, an aquaponic greenhouse is best kept warm for some of the day and should *never* freeze or overheat. If the air temperature drops below freezing for a period of time, the bacteria population will likely die. Then you will have to re-establish the bacteria in a process known as *cycling*, which can take 4–6 weeks. It can be hastened with various methods, but cycling generally takes time, effort and careful monitoring.

As you can tell, there is a lot more on the line with an aquaponics system. That makes temperature control even more important than in a standard greenhouse. Aquaponic greenhouses should take advantage of energy-efficient design as much as possible, and they almost always need to include backup heating/cooling systems.

To provide sustainable water and space heating, we recommend installing a solar hot water system if your location has a good solar resource. This creates a two-in-one solution that can effectively heat both

Case Study: The GrowHaus and Flourish Farms
The Intersection of Farming and Community

**20,000 sq. ft. urban farm and community center
Denver, Colorado**

The GrowHaus is a 20,000 sq. ft. reclaimed greenhouse space in a north Denver neighborhood considered a food desert and one of the most polluted zip codes in Colorado. Surrounded by oil refineries and rail yards, the GrowHaus stands out as a vibrant community center, full of life, educational opportunities and healthy food production. Co-founders Adam Brock and Coby Gould call it a "food hub"—a place where community and food production intertwine. The diverse space includes an urban farm, education and community space, a permaculture demonstration area, and a marketplace that sells affordable and nutritious food in an area otherwise devoid of healthy food options.

The project began in 2009 as a vision for an urban farm centered on food justice. Brock and Gould renovated an abandoned metal greenhouse, replacing the roofing with twin-wall polycarbonate and turning it into a functioning year-round garden. They built out a 5,000 sq. ft. hydroponic growing area, which produces 1,000–1,500 heads of Bibb lettuce each week using nutrient film technique (NFT).

In 2012, GrowHaus partnered with JD and Tawnya Sawyer, owners of Colorado Aquaponics, to integrate a commercial aquaponics greenhouse into the structure. The resulting 3,000 sq. ft. aquaponic space is called Flourish Farms and operated independently by the Sawyers.

Flourish Farms is considered a "hybrid" aquaponics system, combining media beds and deep water culture, as well as wicking beds, microgreen trays and vertical towers. The multiple growing systems maximize production and crop diversity: over 300 varieties of fruits, vegetables and culinary greens have been grown. The most profitable production is the deep water culture the fish tanks and airspace, as described in the Sage School case study, in Chapter 14. Other growers have integrated rocket mass heaters or wood stoves as sustainable alternatives. You will likely want to have an automated backup heating method, such as propane.

Aquaponic greenhouses have higher electric loads and require more outlets than standard greenhouses. Plan out the placement of equipment and outlets during the design phase so you can create a suitable wiring setup. A final consideration is power outages. Fish can't survive

raft system, which produces 800–1,200 heads of lettuce and other greens per week. A portion of the food grown on-site is sold or donated on-site through GrowHaus's small market. The rest of the produce is delivered to local catering companies and restaurants. Using their many years of research and development in aquaponics systems, the Sawyers created a farm that produces zero waste, consumes very few natural resources, and grows over 20,000 lbs. of produce every year.

Colorado Aquaponics hosts workshops for home growers as well as community-scale production managers. GrowHaus offers a range of healthy living classes, from mycology to microbusiness. More information can be found at: thegrowhaus. org and coloradoaquaponics.com.

FIGURE 18.7. Flourish Farms.

long if water oxygen levels drop or if water is not being filtered. If you live in an area with frequent power outages, a small battery backup would be wise. (See Chapter 16 for recommendations.)

Takeaways

- Aquaponics is the growing of fish and plants in a symbiotic ecosystem. It requires a controlled environment that keeps fish, bacteria and plants in a suitable temperature range. An aquaponic greenhouse should be kept warm (>75°F [24°C]) for some part of the day and never freeze or overheat.
- Water tank heating is the #1 energy expense of an aquaponics greenhouse in cold climates. To reduce energy costs, insulate tanks and consider burying them underground.
- Plan aquaponics into your floor plan *early*. Tanks and equipment can occupy a great deal of space in the greenhouse.
- Hydroponics and aquaponics have the potential to maximize yields and reduce water use. They both require research and experience to successfully be implemented in the greenhouse.

Further Reading
Bernstein, Sylvia. *Aquaponic Gardening*. New Society Publishers, 2011.
Ceres Greenhouse Solutions 9 *Steps to Designing an Energy-Efficient Aquaponic Greenhouse*. E-book. Download at ceresgs.com

Endnote
1. Calvin, Linda and Roberta Cook, "North American Greenhouse Tomatoes Emerge as a Major Market Force," April 2005, ers.usda.gov

Temperature Ranges of Common Greenhouse Crops

■ Temperature Range Tolerated
■ Ideal Temperature Range
✗ Ideal Germination Temperature

		28 30 40 50 60 70 80 90 100
Hardy Vegetables (can tolerate hard frost down to 28F)		
Broccoli	Does not tolerate heat (premature bolting)	
Brussels sprouts	Slow growth; tall growth, takes advantage of vertical spaces	
Cabbage		
Collards		
Kale	Does not tolerate heat well; wilting or bolting	
Kohlrabi		
Onions		
Spinach	Bolts in high heat	
Lettuce	Tolerates a range of conditions depending on the variety	
Semi-hardy Vegetables (can tolerate light frost, 30–32F)		
Peas	Good vertical growers for trellising on the North wall	
Beets		
Carrots		
Swiss chard	Tolerates both high and low temperatures; excellent producers for solar greenhouses	
Arugula	Fast growth, can bolt in high heat. Look for heat tolerant varieties	
Parsnips		
Warm Season Vegetables		
Beans	Bush beans or vining; consider vining for tall areas or as an outdoor trellis to provide summer shade	
Celery		

28 30   40   50   60   70   80   90   100

Cucumbers	Excellent warm weather producers; can tolerate heat. Prefer 70-90 temperatures and germinate in soil around 85. Shane Smith recommends the European seedless cucumbers, also called forcing cucumbers, which don't need pollination
Summer squash	

Tender Vegetables

Eggplant	
Peppers	Excellent summer producers because many varieties are heat tolerant
Winter squash	
Tomato	Vining tomatoes grown in beds, and a good way to use vertical space; bush tomatoes grown in beds or containers

Fruit Trees — Most fruit trees are heat lovers, but mature trees are able to survive cold temperatures, down to freezing or sub-freezing for short periods, though they will experience frost-damage. Some, like figs, require cooler temperatures to produce. They require fertile soils.

Banana "Dwarf" Cavendish	Dwarf varieties grow in the range of 6' tall
Meyer Lemon	Lemon trees tolerate short deep freezes
Avocado "Little Cado"	Dwarf variety that grows up to 10' tall
Fig "Black Mission"	Mature trees are very cold-hardy, surviving down to 15F in a dormant phase. They produce fruit in warm temperatures, above 80F.
Dwarf Mandarin Orange	A large variety of dwarf mandarins and tangerines possible
Common Guava	

Herbs & Perennials

Alyssum	Excellent plants to attract / maintain beneficial insects
Sorrel	
Pineapple	Not space efficient, but has a "wow" factor
Rosemary	A hardy perennial, rosemary grows robustly in greenhouses often turning into a bush. It can be pruned and stalks stored (freeze or dry)
Oregano	Grows best in cool temperatures

Optimizing Glazing Angles

This simple process identifies the angle of glazing (in either the walls or roof) that will maximize light transmission for a particular time of year. You can also find the *range* of angles that will allow for sufficient light transmission, taking into account that multiple angles yield nearly equivalent light transmission levels (per discussion in Chapter 5, "Calculating the Angle of Glazing").

First, determine the angle of the sun during the season you want to grow. Ideally, this would be the average angle over the whole season. For instance, you may aim to maximize heat and light during the winter, rather than at winter solstice specifically. You can find the average solar altitude angle for a season by using online tools such as suncalc.org, or see the sidebar in Chapter 4, "Tools for Siting the Greenhouse." Here we'll use an example angle of 30 to indicate the average angle during the winter season at our greenhouse site.

1. First draw your ground plane and sketch the angle of the sun.
2. Draw a line (#1) at a right angle to the angle of the sun. This is your "perfect roof pitch," the perpendicular plane that would receive the maximum light transmission—although it's one that would probably be difficult to build. Keep in mind that this is a representation of a 0-degree angle of incidence, or "normal."
3. Now, draw a second line (#2) 45 degrees from line #1. This represents how far the angle of incidence can deviate from the ideal and yet still yield sufficient light levels; this can be much more practical to build. (It is based on the fact that up to an angle of incidence of 45 degrees, there is little change in light transmission, as shown in Fig. 5.9.) Keep in mind that building to line #2 reduces the area that receives incident solar radiation (the view window) during the winter, so you would want to augment with enough vertical or near-vertical glazing on the walls of the greenhouse. Anywhere between line #1 and line #2 should be a suitable roof slope to ensure enough light during the winter.

4. Finally, you can calculate for the angle of the glazing relative to the ground simply by knowing that all the angles shown add up to 180 degrees. Since three of them are given, simply subtract the sum of the three from 180. In this case the result is 15 degrees for the minimum glazing angle, also considered the roof slope.

#1
0° angle of incidence;
maximum light transmission

#2
45° angle of incidence;
sufficient light transmission

45°

90°

solar elevation angle 30°

15°   angle of glazing from ground

ground

Supplemental Lighting

The most common supplemental lights used for greenhouse growing—called "lamps"—are: fluorescents; HIDs (high intensity discharge); and LEDs (light-emitting diodes). They vary according to their light quality (the wavelengths they produce); light intensity (usually rated in Ft-candles); efficiency (how much light the lamp produces for every unit of energy it consumes in watts); lifespan; and upfront cost. A brief overview of each is given here.

The best use of supplemental lighting is for extending the photoperiod for plants in the winter months. Turning lamps on for a few hours in the evening gives plants a boost of light while still taking advantage of free sunlight as much as possible.

The type of lighting and the duration you use your lights depends on what you are growing. We recommend only using lights if necessary for a few (2–5) hours at the end of the day to extend the photoperiod in the winter. Many growers only use lights for starts because the lamps can be conveniently placed directly over them.

Compact Fluorescents

Compact fluorescents (CFLs) are the bulbs we typically use in our homes. As grow lights, they are typically called T5s. The T stands for tubular—they are long skinny bulbs—and the number represents the diameter of the tube. Thus, as the number increases, so does bulb size. T5s are the latest and most efficient of the fluorescent bulbs (the previous versions being T8s and T12s).

The primary advantage of fluorescents is that they are have a low upfront cost and are widely available. However, they typically have lower light intensities and lower efficiencies than other types of lighting. That makes them ok for home growers who just want to supplement some light, but they're not viable for most commercial growers. Other downsides are the light fixtures that house them; these tend to be large, casting unwanted shade during the day in a greenhouse. This is why many growers use them only to start plants. Starts need lower light intensities and the fixtures (which also don't produce as much heat as other types) can be placed very close to them. Finally, the bulbs decline in efficiency after a few months or a year of use, and need to be replaced, making fluorescents more expensive over the long-term.

High Intensity Discharge (HID)

As the name implies, HID lights produce very high intensities of light. They are a larger investment—typically used by commercial growers who rely on artificial lighting. HIDs can be subdivided into two categories: *high pressure sodium* (HPS) and *metal halide*. HPS lamps produce a yellow/orange light, while metal halide lights produce a broad-spectrum light (that looks bluish or white) which is less useful for photosynthesis, but easier on the eyes. For supplemental lighting, growers most commonly use HPS lamps, unless metal halide is needed (in areas meant primarily for humans to work, for example).

To put HIDs in perspective, one 400-watt HPS lamp produces the same light intensity as ten 54-watt T4 lamps. We rarely see them used by residential growers, except for certain crops that require high light levels.

HID lights come with several extra components, including a ballast to power the light, a reflector to direct light toward the plants, and a fan to reduce heat above the plants.

LEDs

Light-emitting diodes (LEDs) are the newcomers to the market. Though currently more expensive than the options above, many types offer much greater efficiencies (they use less energy to produce equivalent light intensities). There are several misconceptions about LEDs, one of which is that they are lower light intensity than HID lamps. The truth is that they can produce equivalent light intensities but typically cost much more as of this writing. Though it varies by the specific lamp type, as a category, they are more efficient than HIDs.

One drawback with LEDs is the potential for "hot-spots"—concentrations of intense light at the center of the bed. The upside is that, unlike HID lamps, they produce almost no waste heat, so there is no need for fans or vents. We recommend talking to a supplier of LEDs and requesting a lighting layout for your specific growing area. If you are trying to reduce energy consumption as much as possible in your greenhouse, evaluate the upfront cost of LEDs as well as their potential savings over the long-term. If you are considering powering the greenhouse with solar panels, this is even more important.

Index

Page numbers in *italics* indicate tables.